Insect Pests of Cereals and Their Management

Insect Pests of Cereals and Their Management

Sandhya Aggarwal

Oxford Book Company
Jaipur ▪ New Delhi

Published By

Oxford Book Company

Regd. Off. "Bony Residency", Gate No. 2
Opp. Tilak Public School, Vishweswariya Nagar,
Gopalpura Road, Jaipur - 302018, India
Ph: 0141 - 2761280, Telefax: 2761381
email - oxfordbookjpr@gmail.com
website - oxfordbookcompany.com

Delhi Office: 102, 1st Floor, Satyam House,
4327/3, Ansari Road, Darya Ganj,
New Delhi - 110002, India
Ph: 011-45652440

Distributed By
KEYA INTERNATIONAL PRESS LLP
E-mail: keyainternationalpressllp@gmail.com

Published in India, 2024

ISBN : 9788189473822

Price: ₹ 1950.00

Typeset by Shivangi Computers, Jaipur

Printed and bound in India by Thomson Press Ltd.,
New Delhi

Preface

As cereals like rice, wheat and maize constitute a major portion of the world's food produce, apart from being the staple did of many a nation, their agricultural management becomes of the utmost importance with depleting food reserves and food production increasingly becoming a reality no stone can be left unturned in ensuring that cereals are not harmed in either their quality or quantity. One such major factor affecting cereal production is the presence of pest, particularly insects. Annual losses due to pests amount to 20%-40% of potential production of cereals worldwide, an appalling figure considering the need for greater protection of cereals crops in the contemporary world. As such the central of insect pests of cereals becomes even more urgent.

The present text has been designed keeping in mind the need for a manual which provides farmers with the wherewithal to bring insect pests of their cereal crops under management. It brings an extensive knowledge of the pesticides, trends, practices and techniques, both conventional and modem which can be used to combat this menace. In addition, it takes care to incorporate a discussion of specific pest problems, and how they can be managed, with due regard to environmental stability. Further, it also brings in a study of the biological ethics behind using pest-control tactics.

Sandhya Aggarwal

Contents

1

Insect Pests of Rice

On a global basis sufficient food is produced to adequately feed everyone. Yet, the pains of hunger continue to be a common experience of many people in the world today. Food shortages in developing countries are aggravated by rapid population growth. In the last 40 years the increase in the world's population has equalled the increase that occurred from the emergence of the human race to the middle of the 20th century.

The desire for food security has adversely affected the environment. Of additional concern is the evidence of a yield growth rate stagnation in countries where the "Green Revolution" has had its greatest impact. If recent growth rates for cereal demand continue to 2025, the food need in Sub-Saharan Africa will be 2.5 times greater than production. Among the major cereals, rice is the primary staple of more than two billion people in Asia and hundreds of millions of people in Africa and Latin America. Consumption per capita varies greatly from 186 kg/year in Burma to 4 kg/year in the USA.

Today, rice is just as important to food security, or more so, than it was in 1979 when the first Director General of the International Rice Research Institute (IRRI), R. F. Chandler, made the above statement. To cope with the increasing demand for rice a key element is the development and implementation of effective rice insect management strategies.

CLASSIFICATION OF RICE

Rice, an annual grass (Gramineae), belongs to the genus *Oryza* which includes twenty wild species and two cultivated species, *O. sativa* (Asian rice) and *O. glaberrima* African rice). *Oryza sativa* is the most commonly grown species throughout the world today. In Asia *O. sativa* is differentiated into three subspecies based on geographic conditions; indica, javanica, and japonica. Indica refers to the tropical and subtropical varieties grown throughout South and Southeast Asia and southern China. Javanica designates the bulu (awned) and gundil (awnless) rices with long panicles and bold grains growing alongside of indicas in Indonesia. Japonica refers to the short and roundish grained varieties of the temperate zones of Japan, China, and Korea. Japonica-type varieties are grown in northern California, USA due to their tolerance to low night temperatures. Indica-type varieties are grown in the southern USA.

ORIGIN AND SPREAD OF RICE

Domestication of *O. sativa* occurred some 10,000 years ago in the river valleys of South and Southeast Asia and China. Rice specimens found in China date back to 3,000 B. C. and earliest historical Chinese writings indicate that of the five principal food plants in the country, rice was the most important.

Rice was introduced into southern Japan from China in about 100 B. C. and from there it spread reaching the extreme north of Japan only in the eighteenth century. The Portuguese introduced rice into Brazil and the Spaniards introduced rice into Central America and parts of South America.

Rice cultivation in the USA dates from about 1646 when it was introduced into the James River region of Virginia and in 1685 when it was first grown in the colony of South Carolina. The rice variety, "Carolina Gold" was introduced into South Carolina when a storm forced a New England ship, sailing from Madagascar, to harbor in Charleston. Before leaving the port, the captain gave the colonists about 5 kg of rice seed and that started the Carolina rice industry. Rice was introduced into Louisiana in 1718 but did not assume importance there until 1887. Commercial rice production in the Sacramento Valley, California began in 1912.

GROWTH OF RICE PLANT

The parts of a mature rice plant are illustrated in Figure 1. The rice plant consists of the roots, stem, leaves and panicle. Rice passes through the following 10 stages during its growth cycle: (1) germination and emergence, (2) seedling, (3) tillering, (4) stem elongation, (5) panicle initiation, (6) panicle development, (7) flowering, (8) milk grain, (9) dough grain, and (10) mature grain stage. Traditional varieties require about 150 days of growth to reach the mature grain stage whereas the modern, high yielding, very early maturing varieties can be harvested in as few as 90 days after sowing.

Table 1: Rice area, production and yield by world region.

Region	*Area harvested (000 ha)*	*Production (000 metric ton)*	*Grain yield/ ha(metric ton)*
South America	5,659	15,295	2.70
USA	1,336	8,972	6.72
Central America	552	1,929	3.49
Europe	378	2,113	5.59
Asia	130,027	485,077	3.73
Africa	7,235	15,855	2.19
World	1,456,187	529,241	

Rice is cultivated from 53° N to 40° S latitude where it is adapted to a wide range of environmental conditions, from uplands to waterlogged lowlands. It is grown at high elevations, such as the rice terraces in the mountains in the Philippines, and at low elevations down to sea level.

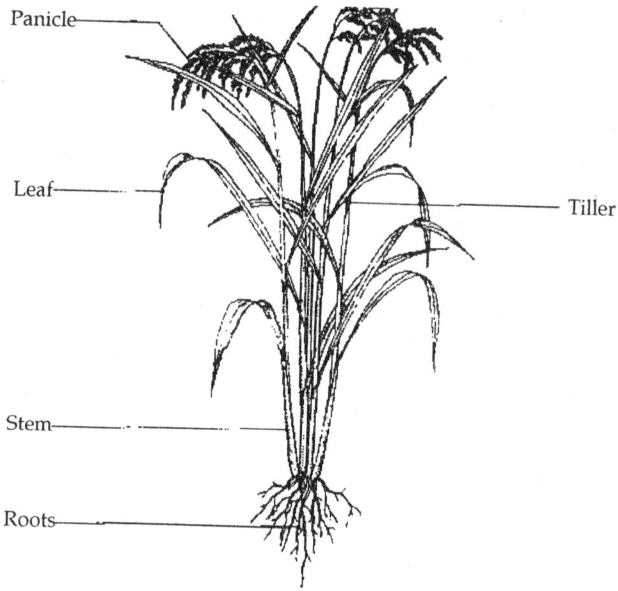

Figure 1: The rice plant

Most of the worlds' rice is grown in Asia with 130 million hectares (ha), whereas there is 1.34 million ha of rice grown in the USA (Table 1). In the USA, rice is grown in Arkansas, California, Louisiana, Mississippi, Missouri, and Texas. Grain yields in the USA are 6.72 m ton/ha whereas they are only 2.19 ton/ha in Africa, due to numerous social, abiotic and biotic constraints, including insects.

RICE PRODUCTION SYSTEMS

There are two major systems of rice cultivation; the dry or

upland system, in which the crop is grown on dry ground, the same as for wheat and other cereals; and the wet system, in which the rice land is flooded and the crop is grown in standing water from planting until near harvesting. Rice production systems more specifically classified according to ecology in terms of water are (1) upland, (2) irrigated, (3) rainfed lowland, and (4) deep water.

Upland rice is often grown in hilly areas with natural rainfall and without levees to impound water. An assured rainfall over a 3 to 4 month period is necessary as the crop is rainfed and the water supply is not controlled.

In the forest region of West Africa, prior to sowing of rice, trees, shrubs and weeds in fallow upland fields are cut and then burned. For rice plants to compete with weeds, rice seed is sown within a few days after burning. Seed is sown directly into untilled land with the aid of a hand hoe.

Irrigated rice is grown in fields with levees to impound water supplied principally from irrigation. Water control is more accurate than with rainfed rice and seldom exceeds 50 cm. Rainfed lowland rice is grown in standing water in fields with levees to impound natural rainfall. If rainfall is heavy, water may reach depths of 50 to 100 cm. Deep water rice is grown in the flood plains along rivers in Mali in West Africa, and India, Bangladesh and Thailand in Asia. Water depth may reach several meters during monsoon rains and varieties adapted to deepwater conditions can elongate at the rate of 25 cm per day in rapidly rising water.

Irrigated, rainfed lowland and deepwater rice fields are ploughed when the rainy season begins and puddled to provide a fine tilth before flooding. Seed may be either broadcast or drilled into the soil or it may be sown in nursery beds and the seedlings subsequently transplanted when they reach a suitable stage of growth (about 21 days after sowing). Rice in the USA is pregerminated and broadcast into irrigated rice fields from airplanes or drill-seeded.

Most of the rice area in China and the USA is irrigated. In Latin America and Africa most of the rice is upland.

Plant Damage

Insect pests attack all portions of the rice plant and all stages of plant growth. Feeding guilds consist of the (1) root feeders, (2) stem borers, (3) leafhoppers and planthoppers, (4) defoliators, and (5) grain sucking insects. Insects also attack rice grains in storage.

Root feeders

Examples of root feeders are termites (order Isoptera) and the rice water weevil, *Lissorhoptrus oryzophilus* (order Coleoptera). Termites, in West African upland rice fields, occur in patches and often kill the plants, especially when rainfall is lacking. The water weevil is a major insect pest of irrigated rice in the USA. The adult water weevil feeds on the leaves and causes little damage while the larvae feed on the roots and severely reduce the root system. Plants with reduced root systems grow poorly and have low yields.

Stem borers

Stem borers consist primarily of insects in the lepidopterous families, Noctuidae and Pyralidae. The adult moths lay eggs on rice leaves and the larvae bore into the stem. Feeding in the stem during the vegetative growth stage of the plant (seedling to stem elongation) causes death of the central shoot ("deadheart"). Damaged shoots do not produce a panicle, and thus, produce no grain. Feeding of stem borers during the reproductive stage (panicle initiation to milk grain) causes a severing of the developing panicle at its base. As a result, the panicle is unfilled and whitish in color, rather than filled with grain and brownish in color. Such empty panicles are called "whiteheads". In the southern USA the sugarcane borer, *Diatraea saccharalis*, and the rice stalk borer, *Chilo plejadellus* are common stem borers attacking rice.

Leafhoppers and Planthoppers

In general, the leafhoppers (family Cicadellidae) attack all aerial parts of the plant whereas the planthoppers (family Delphacidae) attack the basal portions (stems).

The leafhoppers and planthoppers (order Hemiptera) are sucking insects which remove plant sap from the xylem and phloem tissues of the plant. Severely damaged plants dry and take on the brownish appearance of plants that have been damaged by fire. Hence, hopper damage is called "hopper burn".These insects are severe pests in Asia where they not only cause direct damage, by removing plant sap, but are also vectors of serious rice virus diseases, such as rice tungro virus transmitted by the green leafhopper, *Nephotettix virescens*, and grassy stunt virus transmitted by the brown planthopper, *Nilaparvata lugens*.

Defoliators

A large group of insects belonging to several insect orders feed on rice leaves. Most common are the larvae and adults of beetles (order Coleoptera), larvae of the order Lepidoptera and grasshoppers (order Orthoptera). Defoliation reduces the photosynthetic capacity of the rice plant and thereby decreases yields. However, when feeding damage occurs early in rice growth, plants have an ability to compensate for damage by producing new tillers. Thus, rice plants in the actively tillering stage of growth can tolerate a certain level of leaf damage without any yield loss. An example of a coleopterous defoliator, is the adult of the rice water weevil in the USA. Foliage removal by this beetle, however, is usually minimal and below the yield reducing level. The common armyworm, *Pseudaletia unipuncta* (order Lepidoptera) occurs sporadically in epidemic numbers on rice in California where leaf feeding by the larvae causes yield reductions.

Grain sucking insects

The stink bugs (order Hemiptera), known for the foul odor produced by the scent glands on their abdomen, penetrate the developing grain with their sucking mouthparts and remove the white fluid referred to as "milk". Damage early in the development of the grain prevents the filling of the grain. Later attack results in "pecky rice" which is referred to as the condition of the grain after being sucked by stink bugs and the grain being subsequently stained by the bacteria or fungi which enter the puncture wounds. In some countries the market price of pecky rice is reduced. The stink bug, *Oebalus pugnax*, found in North America east of the Rocky Mountains and as far north as Minnesota, is a pest of rice in the southern USA.

Yield Losses

Historical records from Korea and Japan report on the history of rice insect pests during the last two millennia. In 875 a migratory locust outbreak occurred in the Ise District of Japan and a brown planthopper outbreak in 1733 was reported as one of the most damaging insect outbreaks in the history of rice production in Japan. About 2.6 million persons were affected and 12,000 died from hunger.

Insects reduce yields substantially, especially in tropical Asia. Cramer estimated the rice yield losses caused by insects by reviewing the literature up to 1966 (Table 2). He estimated losses ranging from 31.5% in Asia to 2% in Europe. More recently, a series of experiments coordinated by IRRI and conducted in farmers' fields in six Asian countries indicated that the increase in yields of fields receiving fertiliser, and insect and weed control was 0.9 t/ha in the wet season and 1.7 t/ha in the dry season; 0.4 ton and 0.6 ton of these yield increases respectively, were due to insect control.

Table 2: Estimated rice yield losses caused by insect pests on a world basis

Region	*Yield loss (%)*
Asia	31.5
People's Republic of China	15.0
Africa	14.4
South America	3.5
North and Central America	3.4
Europe	2.0

The importance of insects and other pests is indicated by the fact that only a 10% increase or decrease in food grain production, on a global scale, can make the difference between a glut and acute scarcity.

Factors Affecting Extent of Insect-Caused Yield Losses

Although rice insects have been a problem through the centuries, outbreaks have increased and the insect pest complex has changed, in the last four decades. Some insects have increased in severity, whereas others have declined in importance. There is evidence that ecologically specialised (monophagous) species have been favored by crop intensification. Intensification involves changes in cultural practices such as (1) an increase in the number of crops grown per year, (2) an increase in the use of agricultural chemicals (fertiliser and pesticides), (3) increased area under irrigation, and (4) increased plant densities.

Changes in rice cultural practices have accompanied the widespread adoption of modern varieties. The photoperiod insensitivity and reduced growth duration of the modern varieties have made it possible to grow two and even three crops per year, where water, and temperatures are adequate. Continuous cropping throughout the year has caused shifts in the composition of pest fauna. Species dependent on standing water, such as the whorl maggot, *Hydrellia philippina*, and the rice caseworm *Nymphula depunctalis*, have become more abundant because of increased area under irrigation.

The most dramatic effects of modern cultural practices on abundance of rice insect specialists in Asia is seen in the plant sucking Homoptera, especially the nearly monophagous brown planthopper. The brown planthopper arose from the status of a secondary pest to a major yield constraint beginning in the 1960s. The increased use of resurgence-inducing insecticides, has been considered the major cause of intensified brown planthopper problems. Resurgence-inducing insecticides are selectively toxic to the predators of the brown planthopper and result in a dramatic population increase of the brown planthopper after insecticide application.

DEVELOPMENT OF PEST MANAGEMENT TACTICS

As rice scientists and farmers have gained experience in the cultivation of the modern varieties and the agronomic practices that have accompanied the "Green Revolution" there has been a shift from a primarily unilateral approach of insect control, with a strong reliance on insecticides, to a multilateral approach involving a mix of control tactics. This approach, known as integrated pest management (IPM), in the simplest terms is referred to as "a broad ecological attack combining several tactics including biological, chemical, and cultural control methods and insect resistant rice varieties, for the economic control and management of pest populations".

IPM programs have a significant impact on minimising the adverse effects of insecticides, and in increasing the profitability of rice production. It has been estimated that the cost savings from research leading to increased insect pest management efficiency on rice in South and Southeast

Asia will be $973 million in insecticide saved by the year 2,000.

During a 1960s visit to IRRI in the Philippines Norman Cousins, then editor of the Saturday Review, wrote that "any genuine improvement in the human condition on this planet must be concerned with rice." Scientists throughout the rice growing regions of the world are striving to do their part in improving the quality of life by developing and implementing strategies to more effectively and economically manage rice insect pests. These strategies will now be discussed.

Cultural Practices

Cultural methods to control insects involve crop production practices that have a dual purpose of crop production and insect suppression. Farmers have developed these practices through many years of trial and error and these practices have been handed down through generations. Primary cultural control practices are those done specifically to control insects such as draining a field to control the aquatic caseworm larva or planting a trap crop for stem borers. Secondary practices are those that are specifically done for crop husbandry, such as land preparation and weeding, but which also happen to minimise pest buildup.

Rice culture has a rich source of folklore regarding indigenous cultural practices. Cultural control practices that offer potential control of rice insects includes (1) mixed cropping, (2) planting methods (transplanting vs direct seeding), (3) age of seedlings at time of transplanting, (4) water management, (5) fertiliser management, (6) crop rotation, (7) number of rice crops per year, (8) planting time, (9) synchronous vs asynchronous planting over a given area, (10) trap crop, (11) tillage, (12) weeding and (13) growth duration of the crop.

Chemical Control

When modern rice varieties were introduced in the 1960's, insecticides were one component in a package of technologies of the "Green Revolution" in Asia. Chlorinated hydrocarbons were first used, then phosphates, and recently the carbamates have been used. However, insecticides are often too expensive for resource-poor farmers to use. Even where farmers can afford insecticides the health hazards may outweigh the economic benefits. Studies in the Philippines have indicated that when the effect on farmers health and subsequent lost days of work, and the adverse effect on the environment are taken into consideration, the productivity increases obtained from the utilisation of insecticides are minimal. Due to their cost, toxicity to man and the environment, the secondary pest problems caused such as the resurgence of the brown planthopper, and because of the development of insecticide resistant populations, the recent trend in rice IPM has been toward the integration of insect resistant varieties with the conservation of natural control agents. Although there are cases where the judicious use of selective insecticides in rice is necessary, routine, calendar-based applications in a non-IPM context are no longer recommended.

Biological Control

The action of indigenous predators, parasitoids and insect pathogens forms the cornerstone for modern IPM programs on rice. Although hundreds of insect species feed on rice only 8% are considered as major pests. This indicates that over the thousands of years of rice cultivation a relatively stable association between rice insects and their natural enemies has evolved. When this stability is upset, as is the case in the destruction of predators by insecticides, insect outbreaks occur.

Classical and inundative biological control approaches tried so far have had little success in rice. However, research studies have shown that indigenous natural enemies have a strong impact on rice pest populations and their conservation is an essential part of rice IPM programs. Many species of predators, parasitoids and pathogens have been shown to attack rice insect pests. However, only one example of each will be mentioned.

Parasitoids

Numerous parasitoids have been reported attacking the eggs, larvae and pupae of the rice leaffolder, *Cnaphalocrocis medinalis* in Asia. Larval parasitism averaged 40% in studies conducted on the IRRI research farm, and in nearby farmers' fields in the Philippines. From 4,904 larvae collected, 15 species of parasitoids, belonging to 9 families, were found.

Predators

Predacious arthropods, including insects and spiders, attack all stages of rice insects. Spiders are abundant in rice fields throughout the world. The wolf spider, *Lycosa pseudoannulata* is probably the most important predator in rice fields in Asia. The diet of the wolf spider depends on the types of insects that are available but leafhoppers and planthoppers are the major prey. *Lycosa* feeds on both hopper nymphs and adults and is considered to be a major regulator of brown planthopper populations. One wolf spider can eat up to 45 hoppers per day.

Pathogens

Pathogens belonging to the fungi, bacteria and virus groups attack rice insect pests and play an important role in the regulation in rice insect pest populations. Among the entomogenous fungi, *Metarhizium anisopliae,* the "green muscardine fungus", is a common insect pathogen. *Metarhizium anisopliae* commonly attacks rice planthoppers, leafhoppers and rice black bugs, *Scotinophara* sp. A dead black bug, covered by white mycelia of *M. anisopliae.*

PLANT RESISTANCE TO INSECTS

Resistant cultivars are sought as the major tactic in an integrated approach to rice insect control. Incorporation of insect resistance into modern varieties is a major objective of most rice breeding programs in developing countries because:

— income per hectare in rice production is relatively low and
— money spent for controls, such as insecticides, significantly lowers profits,
— resistance in rice cultivars is essentially free to the farmer,
— insecticides cause accidental poisoning accidents and pollute the environment,
— land holdings of most rice farmers are small and thus maintaining
— insecticide application equipment is not economical, and
— varietal resistance is generally compatible with other control tactics.

Because of its unique advantages host plant resistance is sought after as a key tactic in the integrated control of rice insect pests in developing countries. The identification of insect resistant rice germplasm and its use in the breeding of high yielding varieties with resistance to multiple stresses has been an integral part of the of the success of the "Green Revolution" and has increased the profitability of rice production, minimised safety risks to farmers, and contributed to a more healthful environment.

Evaluation of Rice Germplasm

It has been estimated that the number of rice cultivars in collection may reach 120,000. The IRRI International Rice Germplasm Center, established in 1961, has about 80,000 accessions of *O. sativa*, 2,408 strains of *O. glaberrima* and 2,214 wild species and species-hybrids. This collection is the world's largest repository for a single crop. To provide extra security for the IRRI collection the institute deposits a duplicate set of seed in the U.S. National Seed Storage Laboratory, Fort Collins, Colorado, USA.

Methods to screen the germplasm collection to identify insect resistant cultivars for use as donors in the breeding programs have been developed for more than 30 rice insect species throughout the world. In Asia and South America major emphasis has been on the leafhoppers, planthoppers and stem borers. In the USA, major emphasis has been on the rice water weevil.

Breeding Insect Resistant Rice Varieties

Breeding for insect resistance has only been a focus of rice development programs since the early 1960s. Although rice breeding programs throughout the world have developed and released commercial varieties with resistance to numerous insect species, emphasis in this section will be primarily on the brown planthopper as a typical example.

All of the major rice producing countries in South and Southeast Asia have breeding programs to control this pest. Rice cultivars resistant to the brown planthopper were first identified at IRRI in 1963. Since 1963 about 50,000 accessions have been tested and more than 400 resistant accessions have been identified.

Seven genes have been identified that impart resistance to the brown planthopper; Bph 1, bph 2, Bph 3, bph 4, bph 5, Bph 6 and bph 7. IR36, once the most widely grown rice variety in the world has multiple resistance to several diseases and insects, including the brown planthopper, good grain quality, and high yield potential.

Insect Resistant Cultivars as a Component in IPM Systems

Insect resistant varieties are being successfully utilised, as a major component in rice IPM. IPM programs utilise biological controls, traditional cultural controls, insecticides and pest resistant varieties to varying degrees. The cornerstone of the Indonesian rice IPM strategy is the use of multiple pest resistant varieties in a mix with natural biological control and other control tactics. Examples of how insect resistant varieties can be integrated with biological, cultural and chemical control are now presented.

Biological Control

Insect resistant varieties may have an adverse effect on natural enemies by reducing prey density, but they are generally considered to be compatible with biocontrol agents. The combination of multiple pest resistance and natural biological control, is especially attractive in developing countries in the tropics where extension services are often not able to train farmers in IPM principles and practices. Pest resistant varieties shift the pest:predator ratio in favor of biological control. In field studies in the Philippines, brown planthopper: spider ratios decreased with an increase in the level of resistance.

Host plant resistance enhances predatory activity in a manner which may be synergistic. Predation rate of the wolf spider increases when the spider preys on brown planthopper adults on resistant cultivars. The mechanism involved is believed to be the restless nature, and subsequent frequent movement of the brown planthoppers on the resistant variety, which exposes them to a greater degree to detection by the spider.

Combinations of host plant resistance and predation by the mirid bug, *Cyrtorhinus lividipennis* have a cumulative effect on the population increase of the green leafhopper, *Nephotettix virescens,* a vector of the dreaded tungro virus. In cage studies the number of leafhoppers only reached 6 on leafhopper resistant IR29, with the predator, and 31 without the predator, while there were 91 and 220 leafhoppers respectively on susceptible IR22.

Cultural Control

Most cultural practices are considered to integrate well with other control tactics in IPM systems, and the use of resistant varieties, combined with good cultural management, is considered to be a powerful tool in managing pests. Examples of integration are resistant varieties plus (1) fertiliser management, and/or (2) pest evasion techniques.

Increased amounts of nitrogen fertiliser has been one of the major components contributing to the high yields of modern crop cultivars. However, high plant nitrogen levels, are generally favorable to pest insect populations. Results of a study indicated that nitrogen fertiliser favors brown planthopper population growth, regardless of the level of resistance in the host plant, but the population increase is least at the higher levels of resistance, such as on highly resistant IR60.

Cultural practices employed to evade pest attack include early maturing varieties which are harvested before the pest reaches damaging populations. Brown planthopper: predator ratios on early maturing rice varieties are lower than those on late maturing varieties. Incorporation of even moderate levels of brown planthopper resistance into early maturing varieties enhances the level of protection against this pest.

Chemical Control

Because of the weakened physiological condition of insects feeding on resistant varieties control of the surviving insects with insecticides is more effective, when pests are on resistant, than when they are on susceptible, varieties. Brown planthopper mortality, when reared on either a moderately resistant (ASD7), or a highly resistant (Sinna Sivappu) rice variety, is higher than when feeding on a susceptible (TN1) variety.

Another interaction of host plant resistance and insecticides is the relationship between level of resistance and insecticide-induced, brown planthopper resurgence. Brown planthopper populations on a resistant rice variety treated with a resurgence-inducing insecticide only reached 10 insects per plant, whereas the population on a treated susceptible variety was 1,100 per plant. Thus, when insects other than the brown planthopper reach populations above the economic threshold, and require insecticide treatment, the level of brown planthopper resurgence can be reduced, by planting a brown planthopper resistant variety.

Development of "Biotypes" on Insect Resistant Rice Varities. Insect populations have a wide range of genetic variability that maximises their fitness in the presence of genetic diversity of host plants. The widespread planting of one rice variety (monocrop) that has been commonplace since the "Green Revolution" has significantly decreased the genetic diversity of rice plants. As a result, some rice insect species have overcome the resistance of certain rice varieties. Such apparently new forms of pests have been termed 'biotypes' which refers to a population of insects that is capable of damaging plant varieties that are resistant to other populations of the same species.

Brown planthopper biotypes have been a very severe problem in South and Southeast Asia. IR26 with the Bph 1 gene for resistance was the first brown planthopper resistant variety released by IRRI in 1974. Within the Philippines, brown planthopper outbreaks were observed in IR26 after 2 to 3 years of commercial cultivation (approximately 6 crops) as the result of a selection of a strain that could feed on IR26 (biotype 2).

To cope with the brown planthopper biotype problem several gene deployment strategies have been proposed to increase the stability of insect resistant varieties. These strategies are (1) sequential release where a variety with a single major resistance (R) gene replaces a variety with an R gene that has been overcome by the selection for a virulent biotype, (2) gene pyramiding, the incorporation of two or more major R genes into the same variety to provide resistance to two or more biotypes, (3) horizontal resistance, a type of resistance that is expressed equally against all biotypes, (4) gene rotation, a strategy where varieties with different R genes are used in different cropping seasons to minimise selection pressure on given resistance genes, and (5) geographical deployment, the planting of varieties with different R genes in adjacent cropping areas.

USE OF BIOTECHNOLOGY

Developing Insect Resistant Rice Varieties. Recent advances in biotechnology provide the possibility of solving some of the constraints that have limited the practical use of genetic resistance to insects in pest management programs. Biotechnology provides new possibilities of manipulating germplasm.

Wide hybridisation. Wide hybridisation is a plant breeding tool for the incorporation of alien genetic variation from wild species of *Oryza* into commercially useful cultivars. Wild rice species are a rich source of R genes for use in breeding programs for insect resistance. The IRRI breeding program has utilised the wild rice species *O. brachyantha* in the development of crosses with *O. sativa* in the development of brown planthopper resistance. Some progeny are resistant to all known biotypes of the brown planthopper.

Genetic Transformation. IRRI scientists in cooperation with the Rockefeller Foundation Biotechnology Network are incorporating novel genes for resistance into rice through transformation. Genes, such as the Bt (*Bacillus thuringiensis*) gene coding for toxic proteins, inhibitors of digestive enzymes such as protease inhibitors, and ribosome inactivating genes are being transferred to rice.

Insect Resistant Cultivars and Control of Non-insect Pests

There are a myriad of interactions among the various pests (insects, diseases, nematodes, rodents, and weeds) that attack rice plants. The control of a target pest may cause population increases, or decreases, or may not affect another pest.

Weeds and plant diseases may be adversely affected by the planting of insect resistant varieties. Insects that feed on susceptible seedlings and plants, in the vegetative stage, slow rice plant growth and reduce the plant foliage area. This causes the closing of the plant canopy to be delayed and reduces the competition for light and nutrients which in turn allows greater weed growth.

Insect feeding predisposes rice plants to attack of bacterial and fungal pathogens that invade the insect-damaged plant tissue. Some insect species, especially the leaf- and planthoppers, are rice virus vectors. Decrease in the feeding activity on non-preferred rice varieties decreases the incidence of virus infection.

Utilisation of Insect Resistant Rice Varieties

Host plant resistance has served as a key component in IPM programs and in the development of sustainable rice production systems. National rice research programs in many countries have made significant progress in the development and release of pest-resistant varieties. It would be difficult to cover in detail all of these programs so we will limit our discussion to the program at IRRI in the Philippines.

IRRI's rice improvement program has placed major emphasis on developing germplasm with multiple resistance to key diseases and insects. The widespread adoption of such varieties has helped to stabilise rice production in Asia. When new varieties with multiple resistance are introduced there are only minor fluctuations in yield, as these varieties have greater yield stability. Yields of the original high-yielding varieties such as IR8 would have declined an average of 1.3% per year, or a total of 2 t/ha over a 20-year period without crop improvement to incorporate multiple pest resistance. The sequential release of varieties with diverse genes is necessary to cope with the development of new races and biotypes of diseases and insects respectively.

Modern varieties have rapidly spread in the major rice growing countries and have had a significant economic and social impact. In the Philippines, 94% of the area is planted to modern varieties, and modern varieties occupy 100% of the rice area in China Japan and the USA. Variety IR36 with multiple resistance to insects and diseases was once grown on about 16 million ha in Asia. The availability of rice varieties with multiple resistance to insects and diseases has stabilised yields, increased farmer income, minimised the need for pesticides and has promoted the adoption of IPM practices.

Weed Management in Small Holder Rice Production in the Tropics

Rice is the most important food crop in developing countries, and accounts for 29% of the total calorie intake of these populations. In the tropics, rice is largely grown on small family farms, which are usually less than 4 ha, and in Asia, rice is the most important source of employment in rural areas. Throughout these smallholder systems, weeds are one of the major biological constraints to production.

RICE CULTIVATION

Rice is grown throughout the tropics in rainfed uplands, seasonally deep flooded areas, and in rainfed and irrigated lowlands. Upland rice describes the cultivation of "rainfed" rice on well drained, non-irrigated fields, and the crop is grown in Asia, Latin America and Africa. Rice grown on seasonally deep flooded areas, "deep water or floating rice", is important in many areas in West Africa and Asia, often along the courses of major rivers. This crop is sown before the flood waters rise and is able to survive the subsequent deep flooding. Lowland rice describes crops where the fields are inundated for at least part of the time between establishment and harvest, and may depend on rainfall or irrigation.

Table 3: World Rice Production.

	Population (mill).	*Share of the rice yield (%)*	*Labour force (mill)*	*Yield (t/ha) in agriculture rice area*
Latin America	445	26	6,267	2.8
Africa	645	63	6,607	2.1
Asia	3,157	60	133,251	3.7
Total	5,350	47	148,366	3.5

Table 3 shows regional rice yields (t/ha) compared to total population and the annual increase in the area cultivated to rice. The average regional yields conceal substantial variations depending upon the production systems and levels of inputs.

Upland Rice

Almost 100 million people depend on upland rice as their

staple, including some of the World's poorest farmers. In West Africa, upland rice comprises 57% (1.8 million ha) of the total rice area. Upland rice is grown in diverse systems, ranging from shifting cultivation to relatively intensive systems, utilising hand, animal or mechanised tillage and rotations with other crops, including cotton, legumes and other cereals. Shifting cultivation occurs throughout the humid forest zone, where land is cleared from forest, usually by slash and burn, and rice is grown for one or more seasons before the land is returned to fallow. Invasion by weeds is a principal reason for abandoning land after periods of cultivation. Upland rice is dibble, broadcast, or row seeded, the former being commonly used in shifting cultivation. The most common method of weed control is by hand, often with the aid of hoes or machetes. In Asia, Africa and parts of Latin America upland rice is typically grown with few, if any, purchased inputs. A wide range of weeds infest upland rice, many of which are pan-tropical, including the grass weeds: *Digitaria* spp., *Echinochloa colona, Eleusine indica, Paspalum* spp., and *Rottboellia cochinchinensis*, and the broadleaf weeds: *Commelina* spp., *Ageratum conyzoides, Portulaca oleracea, Amaranthus* spp. and *Euphorbia* spp. The variability of weed species composition in upland rice tends to be greater than in the other production systems, and is dependent upon ecology, the cropping system and management practice.

Lowland Rice

These systems vary considerably according to the degrees of land levelling, drainage, bunding, to enable either rainfall to be retained or irrigation water supplied to flood the rice crop. The rice crop may either be direct sown or transplanted from a nursery bed. Production systems with effective water control present the greatest technical opportunities for intensification and diversification of the cropping system. In Asia, the majority of land suitable for intensive lowland rice production is already under cultivation. Flooding usually improves soil chemical conditions and the anaerobic soil environment prevents the germination and growth of many weeds. In the direct seeded systems however, where the field cannot be adequately flooded until the crop is established, post-emergent weed control is usually essential. Weed infestations in irrigated rice are frequently exacerbated by poor land levelling and preparation, inadequate water management, irrigation water and rice seed contaminated with weed seeds, direct seeding, and no crop rotation. In West Africa, of the 20-50 million ha of inland valley bottoms and hydromorphic fringes, only about 15% are currently utilised for crop production. Appropriate weed control regimes are a vital component of sustainable development for such areas. The most serious weeds of lowland rice are pan-tropical and the more widespread include sedges, such as *Cyperus difformis, C. iria, Fimbristylis* spp. and *Scirpus maritimus*, the grasses *Echinochloa crus-galli, E. crus-pavonis, E. glaberescens, E. pyramidalis, Ischaemum rugosum, Leptochloa* spp., *Oryza barthii, O. longistaminata, O. rufipogon* and the broadleaves *Ludwigia* spp., *Eclipta prostrata* and *Sphenochlea zeylanica*. These weeds are well adapted to the aquatic environment, capable of rapid growth and multiplication and are very competitive with rice.

The similarity of some weeds to rice, such as *Echinochloa* spp. at early stages of growth, makes it very difficult for farmers to distinguish them while hand weeding. Some lowland weeds favor particular environments, such as the rhizomatous grasses *Paspalum vaginatum* and *P. distichum* in the coastal and inland saline marshes. The distribution to farmers of rice seed contaminated with weed seeds is at least partly to blame for the widespread occurrence of many of the more serious rice weeds. The introduction of weeds such as *Oryza* spp., *Ischaemum rugosum, Echinochloa* spp., *Rottboellia cochinchinensis*, and *Euphorbia heterophylla* into previously un-infested areas has occurred in this way. In the lowland rice systems, weed seeds may also spread rapidly in irrigation and flood waters.

Losses Due to Weeds

In a survey of upland rice producing countries covering 80% of the total production area, weeds were the most widely reported biological constraint to yields. Upland rice, in particular, competes poorly with weeds and uncontrolled weed growth often results in negligible or zero yield. In West Africa, yields of upland rice with farmers' weed control, were 44% lower than on researcher weeded plots. Losses due to uncontrolled weed growth in upland rice in India were up to 90%, and in both lowland and upland systems in Africa losses were within the range 28-100%. In Belize, Central America, two years after clearing a fallow, yields of upland rice with no weed control were less than 20% of those when the crop was hand weeded twice after sowing. Losses can be particularly severe in direct seeded lowland rice, as the rice and weed seedlings are at similar growth stages. In transplanted rice, the young rice plants have an advantage over germinating weeds and immediate flooding after transplanting limits the establishment of many weeds, hence yield losses to weed competition tend to be less than those in direct seeded rice. In Asia, yield losses due to uncontrolled weed growth in direct seeded lowland rice was reported to be between 45-75%, and for transplanted lowland rice approximately 50%. It is rare however, for farmers not to undertake some weed control and therefore losses on farmers' fields are likely to be considerably less.

Weed Control Methods

Integrated management: To a greater extent than in the management of insect pests and diseases, the management of weeds requires integrated strategies to be successful.

While in some production systems, herbicides may provide the main means of control, these alone are unlikely to be successful unless combined with good land preparation and, in the lowland systems, good water control. No one weed control method is likely to control all weeds, and in the long term this can lead to a build-up of certain species. The combination of direct weed control methods, such as herbicides or hand weeding, with indirect methods such as land preparation, flooding and a competitive crop, will help prevent this situation. A combination of direct weed control methods such as herbicide use with subsequent hand weeding can result in improved yields (Table 4). The optimum combination of weed control methods will depend on the farming system, economic conditions, and farmers' resource and knowledge base.

Table 4: Effect of different weed control methods on grain yield in lowland dry seeded rice.

Control method	Grain Yield t ha
Butachlor	2.2
Pendimethalin	0.9
Butachlor + hand weeding	3.9
Pendimethalin + hand weeding	2.8
Propanil + MCP	3.1

Land preparation : Tillage serves to provide a suitable soil tilth for a seed-bed and control weeds prior to crop establishment. In smallholder systems, practice varies from zero tillage, as in many of the systems of shifting cultivation, to repeated deep cultivation to remove troublesome perennial weeds, such as *Oryza longistaminata* which has an extensive rhizome system. Shallow tillage is often ineffective in controlling weeds and, regardless of tillage practice, post-emergence weed control is normally necessary.

Flooding : The flooding of fields is the most effective method of cultural control of weeds in rice. Flooding to a depth of 10 cm prevents germination of most weed seeds and kills the majority of weed seedlings (Table 5). Normally, flooding is used in conjunction with other control measures, such as herbicides or hand weeding. However, for flooding to be successful, water levels must be maintained and fields well levelled to ensure an even depth of water. In many smallholder schemes, limited irrigation water and poor land development can be major constraints to effective weed control.

Table 5: Effect of water depth on weed growth in rice.

Water depth cm	Weed weight g m^{-2}
57	2.5
20	7.5
11	12.5
8	17.5

Hand weeding : Hand weeding is the most widely used weed control method, with availability of labour being the main limitation to its effectiveneşs. In some areas, adoption of line planting in transplanted rice has allowed the introduction of rotary weeders for cultivation between rice rows, considerably reducing labour requirements for weed control. In India, hand weeding rice at 15 and 30 DAS gave a 60% increase in yield over a single hand weeding at 30 DAS. However, because hand weeding is laborious, with labour sometimes expensive and in short supply, weed control is often imperfect and/or delayed. In a survey of rice farmers in Côte d'Ivoire, 53% said that their fields were not always weeded, with the most common explanation being that weeding was not considered worthwhile due to severe weed infestation, thus effectively abandoning the crop. Further, 80% of farmers said that if weeds were less of a problem they would increase the area of land under cultivation.

Several constraints limit the effective use of hand weeding, including household labour constraints, limited cash for hiring labour, and labour not being available for hire during peak periods. In a village survey in Côte d'Ivoire, where no herbicides were used, farmers spent 408 and 506 hours hand weeding upland and lowland rice, respectively. In Asia, weed control in upland rice has been reported to require 32 - 198 man days per ha, representing 17 - 57% of the total labour requirement for the crop. The availability of animal traction can alleviate the constraints posed by reliance on hand labour. Technology including an animal drawn row seeder and hoes, enabling mechanical weed control, may be an appropriate package where animal traction is a possibility, although problems have been encountered with the operation of seeders under farmer conditions.

Fallow systems : Man's inability to control weeds is one of the major constraints to continuous cultivation in the tropics, and is a particularly acute constraint for those with limited resources and little access to appropriate technologies. After initial clearance of the forest, weed growth consists largely of broadleaf weeds and forest regrowth, though with repeated cropping, more problematic grasses invade the fields. Studies have shown that weed growth following a fallow of less than three years was almost twice that following a fallow of five or more years. Where the cropping cycle is short compared to the length of fallow, forest rapidly regenerates from seeds, roots and cut stumps. However, repeated cultivation allows invasion by annual and perennial weed species and replacement of native forest species by weeds such as *Imperata cylindrica* and *Chromolaena odorata*. While shifting cultivation has been criticized as being wasteful of land and forest resources, it is still the main means of production for many resource-poor farmers. In such systems, the fallow period is an integral part of the cropping cycle. In many areas where shifting cultivation is practiced, the soils are inherently infertile. Cropping intensification on such areas therefore, requires technologies to address increased weed pressure and declining soil fertility, if such systems are to remain sustainable. Research has aimed at improving fallows, often utilising legumes to enhance soil fertility and particularly nitrogen status, reduce weed growth and protect the soil from erosion. Possible legumes species for rice-based systems include the "creeping" species of *Calopogonium*, *Pueraria* and *Mucuna*, and shrub species such as *Sesbania* and *Aeschynomene*.

Crop competition: Rice is not very competitive with weeds during the seedling stages, though this can be an important factor during the vegetative and reproductive stages. Crop competitiveness with weeds is particularly important to limit weed infestation after the initial weed control treatments , such as land preparation, herbicides and hand weeding, have been undertaken. Crop and weed interactions largely involve the competition for light, water and nutrients. Where water and nutrients are not limiting, light becomes the most important limiting factor. Weeds however, usually have higher growth rates and nutrient demands than rice, hence the amount and timing of fertiliser application can significantly effect weed-rice competition.

In an experiment on moist but not flooded soil in West Africa, an application of 50 Kg N ha^{-1} to rice doubled the weed biomass in the crop. Suggesting that in such situations, fertiliser application would have to be accompanied by improved weed management. Reducing the distance between rows or hills, has been shown to reduce weed infestation in rice in upland, hydromorphic and lowland ecologies (Table 6). Likewise, increased weed populations can decrease crop yield loss (Table 7). In irrigated lowland systems in West Africa, crop seed rates up to 200 kg/ha have been reported to be used by farmers to improve weed suppression by the crop. It has been noted that some "modern" higher yielding rice varieties are less able to compete with weeds than many of the traditional rice varieties they were intended to replace.

Traditional varieties tend to be tall, of longer duration, droopy leaves and high vegetative vigor, while the modern varieties often combine short stature, short growth duration and erect leaves. This suggests that increases in yield potential of modern rice varieties compared to tall, leafy traditional varieties has been achieved by sacrificing their competitive ability with weeds. The lack of weed competitiveness in modern varieties may be one reason that many upland rice farmers have retained traditional varieties. The competitive ability of different rice varieties has become a focus of research, with the intention of combining competitive ability with other desirable characteristics in order to develop varieties suitable to low-input conditions. Studies on different plant types show that tillering ability, height, leaf canopy and root development may be important factors in determining the competitive ability of rice plants.

The African rice, *O. glaberrima*, is a source of a number of these traits which confer competitive ability with weeds, and to produce rice plant types which are more suited to smallholder conditions *O. glaberrima* / *O. sativa* hybrids have been produced. Crop varietal development related to weed control also includes host plant resistance to the parasitic weed, *Striga*, and the possibility of utilising crop allelopathic activity. Rice cultivars resistant to *Striga*, a localised problem in some savanna areas of Africa, have been identified and field tested, providing a potential solution for affected areas. Allelopathy is the ability of one plant to interfere with the growth of another by the production of root exudates. Javanica type *O. sativa*, *O. glaberrima*, and wild rices have been suggested as possible sources of allelopathic activity, which could be transferred into commercial rice cultivars. While the utilisation of allelopathy may have considerable potential for low input weed management regimes, the availability of such technology to the farmer is not likely to occur in the near future.

Table 6: Effect of plant spacing on weed weight in lowland irrigated rice, West Africa.

Plant Spacing cm	*Weed weight g m^{-2}*	*Grain Yield t ha*
10	29.0	4.75
14	40.6	4.26
20	69.5	5.01
30	82.3	4.37
40	106.5	3.39

Table 7: Grain yield loss due to competition from Echinochloa sp.

Weed density (plants/m^{-2})	*Yield loss %*
11	25
54	49
269	79

Chemical control: The importance of herbicide use in the tropics is closely related to the cost and availability of labour. Herbicides are one of the first labour saving technologies to be adopted as labour costs rise. As a consequence, the use of herbicides varies considerably between countries. Many rice crops in Japan receive more than two herbicide applications, while in the Philippines only about 50% of the rice area is treated, and in Bangladesh, where rural labor is relatively inexpensive, there is little herbicide use. Herbicides replace hand weeding and enable direct seeding rather than transplanting, which is less labour demanding. Direct seeding in the lowlands is linked to the use of herbicides, as without their use the weeds grow so rapidly in the stages before the fields can be flooded, that manual means of control are often not feasible. Herbicides are also used in the transplanted systems, though to a much lesser extent, and in the upland systems particularly where rice is grown in rotation with cash crops. The costs involved with herbicide use are likely to remain a major constraint to their widespread adoption in many regions. Herbicides may be classified as non-selective or selective, and pre- and post emergence. Most herbicides used in rice production are selective, controlling some or most weeds, while having a limited effect on the crop. Selectivity is not necessarily dependent upon the compounds, but also on the rates, timing and methods of application, and hence it is important to follow the manufacturers recommendations.

Non-selective herbicides such as glyphosate are sometimes used before establishing rice, on weed infestations such as wild rice which are difficult to control with selective herbicides. Pre-emergence herbicides are applied to the soil and control weeds before they emerge, while post-emergence are applied to weeds after they emerge. Among the anilide group are the herbicides butachlor, pretilachlor and propanil. Butachlor can be applied either as pre-emergence or early post emergence to give control over a wide range of annual grasses and some broadleaf weeds.

Pretilachlor can be used safely on transplanted rice, or with direct seeded rice if applied with an antidote. Propanil, a contact herbicide with no residual effects, is a well established herbicide which gives control over a range of grasses and some broadleaf weeds at the 2 to 3 leaf stage, though control is less effective at later stages. Propanil is widely used in proprietary herbicides combinations or in tank mixtures. Elsewhere in the world, such as in Latin America, where propanil has been used intensively some resistance to the herbicide has developed among *Echinochloa* spp. Alkanoic acids (phenoxy acetic acids) include: 2,4-D, 2,4,5-T, MCPA, and triclopyr, and these are foliar applied, translocated herbicides which selectively control broadleaf weeds, sedge and some grass seedlings.

These herbicides have been widely used in proprietary and tank mixtures, particularly with propanil, to give a wide spectrum of control; such mixtures are among the more favorably priced herbicide treatments. Due to the high biological activity of some of these compounds eg. 2,4-D, there is degree of risk attached to their use, as spray drift or contamination of application equipment may affect crops such as cotton and a range of the vegetables. In recent years, 2,4,5-T has been withdrawn from many markets due to concern for human health. Thiocarbamates include molinate and thiobencarb, both soil active herbicides.

The former requires moisture to be active and once applied the area should be continuously flooded; molinate is particularly effective at controlling *Echinochloa* spp. Thiobencarb controls a range of grasses and sedges and may be applied pre- or early post emergence. The dinitroanilines include the pre-emergence herbicides butralin and pendimethalin. These are effective against a range of broad leaf weeds and grasses when applied before the germination of weeds, though they have little effect against established weeds. The latter gives good control over *Rottboellia cocohichinensis* and *Setaria* spp. Diphenyl ethers include bifenox,, flourodifen and oxyflurofen.

These are used as pre-emergence or early post-emergence herbicides and are effective against a range of broadleaf and grass species at the seedling stages. Simetryn and dimethametryn are triazine compounds used as selective herbicides in rice, and are applied at pre or early post emergence. The former may be combined with thiobencarb, while the latter is sold as a proprietary mix with either piperophos or pretilachlor. Bensulphuron-methyl is a herbicide of the sulphonylurea group, and can be used in both seeded and transplanted rice. Applied as pre-emergence or early post emergence, the herbicide gives control over a wide range of broad leaf weeds and sedges. In common with other members of the sulphonyl ureas, this herbicide has a high rate of biological activity, and only very small quantities of the active ingredient are required for effective weed control. Bentazon is a contact herbicide which gives good post emergence control over a range of broad leaf and sedges weeds. Bentazon is marketed either as a sole product or as a mix with propanil. Oxadiason is a pre-emergent herbicide though with limited post emergent action. It controls wide range of weeds and has good persistence in the soil due to low solubility in water and high adsorption in the soil. It is available in various formulations allowing it to applied with conventional sprayers to the soil alone, or with propanil, or using shaker bottles directly to the flood-water in the rice field.

Problems with herbicide application : Smallholder farmers often face a number of problems related to herbicide use, due to either an inadequate rate of herbicide being applied, or the herbicides being applied too late to provide good effect on the weeds. A major cause of this is likely to be the serious lack of information available to the farmer and the poor level of understanding that results. Often only minimal precautions are taken with regard to *health* and *safety* and the use of herbicides. While the majority pesticides applied to agricultural crops are probably insecticides, and that herbicides are usually less toxic to man, it is still a matter of concern.

Poisoning may occur through ingestion of pesticides, skin absorption, or inhalation, and lack of appropriate protective equipment and training make it almost inevitable. Poor standards of pesticide use and ignorance of the possible dangers could be blamed on low levels of education, but poor packaging with inadequate warnings and a lack of information available to the farmer are also major factors. In the early 1970's in Asia, formulations of

2,4-D and MCPA were recommended for controlling annual weeds in transplanted rice, while granular formulations of the selective herbicides butachlor and thiobencarb were reported effective in direct seeded rice, as alternatives to hand weeding.

The subsequent adoption of such herbicides was widespread. Since that time, herbicides have been developed which are safer to use in rice and which have allowed much greater flexibility in application. Formulations allowing the application of herbicides directly to irrigation water without the use of spraying equipment have advantages for the small farmer and have become established practice in many areas. With frequent use of herbicides, herbicide resistance has evolved within some weed populations. In the USA, 30 years of propanil use resulted in resistant *Echinochloa* sp., and after four years of continuous use, bensulfuron resistance emerged in four aquatic weed species.

In Costa Rica, *Ixophorus* sp. and *Eleusine indica* were found to have evolved imazapyr resistance after about five years of herbicide use. Studies suggested that selected biotypes of *Ixophorus* were between 5 to 80 times more resistant to imazapyr than the most susceptible biotypes. The evolution of herbicide resistant weeds is a real threat to effective weed control where herbicides are frequently used. Smallholder systems may be particularly vulnerable as herbicide are often not used at appropriate times or dosages, which may hasten the development of resistance.

REFERENCES

Chang, T. T., *The origin, evolution, cultivation dissemination and diversification of Asian and African rices*, Euphytica, 1976.

De Datta, S. K., *Principles and Practices of Rice Production*, John Wiley, New York, 1981.

Grist, D. H., *Rice*, Fifth ed., Longman, London, 1975.

Heinrichs, E. A., F. G. Medrano and H. R. Rapusas, "Genetic Evaluation for Insect Resistance in Rice", International Rice Research Institute, Los Baños, Philippines, 1985.

Heinrichs, E. A. and T. A. Miller, *Rice Insects: Management Strategies*, Springer-Verlag, New York, 1991.

Shepard, B. M., A. T. Barrion and J. A. Litsinger, "Friends of the Rice Farmer: Helpful Insects, Spiders, and Pathogens", International Rice Research Institute, Los Baños, Philippines, 1987.

Swaminathan, M. S., *Rice*, Scientific American, 1984.

Tribe, D., "Feeding the Greening World", *The role of international agricultural research*, CAB International, Wallingford, UK, 1994.

2

Insects Pests of Wheat

Wheat and triticale can be attacked by a great many insects. Fortunately, only a few insect species are of major importance, causing severe damage over large geographical areas; most species are only occasional pests and/or are not geographically widespread. The "pest status" of many of species is not always well documented.

APHIDS

Symptoms: Aphids are nearly transparent, soft-bodied sucking insects. When present in sufficient numbers, aphids can cause yellowing and premature death of leaves. They exude drops of sugary liquid known as "honeydew", which may cause tiny scorch marks on the foliage and tends to encourage the development of sooty molds. The feeding of *Schizaphis graminin* is especially damaging, resulting in the development of necrotic areas sometimes accompanied by purpling and rolling of the infested leaves. The feeding of *Diuraphis noxis* produces long white stripes on the leaves, leaf rolling, postrate growth habit, and sterile heads.

Life Cycle: The life cycle of aphids involve winged (alates), wingless (apterous), sexual, and asexual forms. When feeding on cereals, the females of most aphid species reproduce asexually (without being fertilised), giving rise to nymphs rather than eggs.

Hosts/Distribution: Species commonly found on cereals throughout the world include:

— *Rhopalosiphum padi* (bird cherry-oat aphid)

— *Schizaphis graminum* (greenbug)

— *R. maidis* (corn leaf aphid)

— *Metopoliphium dirhodum* (rose grass aphid)

— *Sitobion avenae* (English grain aphid)

— *Diuraphis noxis* (Russian wheat aphid)

Importance: Aphids are important and widespread pests on cereal crops. When feeding in sufficient numbers, they can cause significant damage. In addition, the species listed above may act as vectors of barley yellow dwarf virus.

STINK BUGS

Symptoms: Adult stink bugs feed on stem tissue or developing kernels. Saliva from this insect is toxic to the plant, and a single feeding puncture can kill a stem. Feeding on kernels during the milk dough stage will destroy the kernel, while feeding during later development stages will badly shrivel the grain. Feeding on the developing head may cause partial or total sterility. Adult stink bugs have a shield-shaped body and emit a disagreeable odor when crushed.

Life Cycle: Stink bugs over-winter as adults and may diapause. They tend to hibernate under dead leaves and grass. In the spring they migrate to cereal hosts, mate, and lay eggs at various places on the plant. These hatch into nymphs that feed on the plant. Mild winters and low rainfall seem to favor outbreaks of the insects.

Hosts/Distribution: Stink bugs will feed on most cereals and grasses, as well as a large range of weeds (depending on the species). Stink bugs are of major economic importance in Asia Minor.

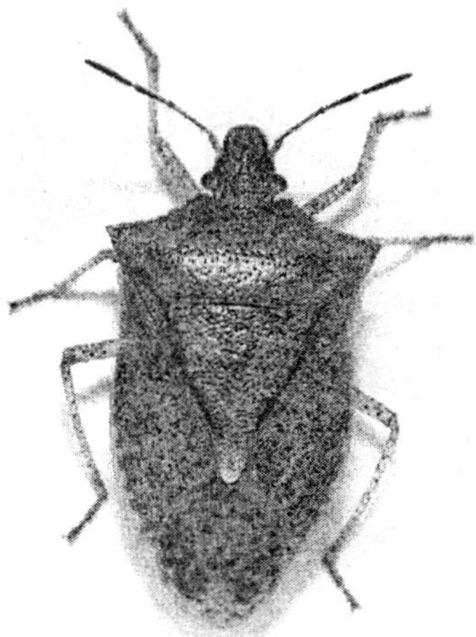

Stink Bugs

Importance: Losses due to stink bugs are highly variable and depend on the density of the insects, weather conditions, and duration of the crop growing period. Losses are due primarily to reduced baking quality.

ARMYWORMS, CUTWORMS, AND STALK BORERS

Symptoms: The primary symptom is defoliation of the plant. Larvae feed on leaves, chewing from the edges to the midrib, or on the heads of cereal plants. Heavy infestations can be very destructive; larvae may climb the plant and sever the neck just below the head. Some species may be found feeding at the soil surface, others underground feeding on roots, and still others feeding inside the stem.

Life Cycle: Adult cutworms and army worms are moths, and the females lay eggs on leaves and leaf sheaths near the ground. These eggs hatch within a few days and initially the larvae feed close to where they hatch. The larvae are found in cracks in the soil or under rocks during the day, feeding at night or early in the morning. In damp weather, they may feed all day.

Armyworms

Hosts/Distribution: Larvae are generally omnivorous in attacking grasses. Species of these insects are found in most cereal-growing areas of the world.

Importance: Cutworms and armyworms sporadically cause severe damage; when they do, they can devastate large areas.

CEREAL LEAF BEETLE

Oulema Melanopa

Symptoms: Adult beetles are 4-5 mm long, have a black head, light brown thorax, and a shiny blue-green wing cover with parallel lines of small dots. Larvae are a dull to bright yellow color, but soon take on the appearance of a slimy, globular, black mass due to the mound of fecal material they produce and accumulate on their backs. The most prominent symptom of cereal leaf beetle infestations is the distinct, longitudinal stripes on leaves; these stripes are produced by the feeding of adult beetles and of larvae.

Life Cycle: The insect produces one generation per year. Adults begin their feeding activity in the spring. They lay yellow eggs, either singly or in small chains, covering them with a sticky film that the soil and the adults emerge in summer. Adults overwinter underneath plant debris on the soil surface, in leaf sheaths and ears of standing maize, or under the bark of trees.

Hosts/Distribution: Cereal leaf beetles can be a problem on fall-sown cereals. Wheats with hairy leaves are affected less.

Importance: Significant yield losses can occur in winter wheat and fall-sown spring wheat. Yield losses of from 14% to more than 25% have occurred with natural infestations.

Thrips

Symptoms: Thrips are small (1 mm long), brown or black insects with a tapering, segmented abdomen. They have piercing and sucking mouthparts and usually have two pairs of narrow wings.

Thrips

They are usually found behind the sheath of the flag leaf, feeding on the stem. However, leaves, stems, and heads may be attacked. Adults and nymphs both can cause damage and, if present in large numbers, may cause the tissue on which they are feeding to take on a silver coloration.

Life Cycle: Eggs are inserted into or attached to host tissue. The generation time is very short, and there may be 10 or more generations per year. Heavy rains will usually destroy a high proportion of they population.

Host/Distribution: Several thrips species live exclusively on cereals, and on forage or weed grasses.

Importance: Thrips rarely cause serious damage, and it is unusual to find infestations at such a level as to warrant control.

Hessian Fly

Symptoms: Severe infestations of Hessian flies result in stunting of the plants, thin stands, lodging, and reduced yield. Injury is caused entirely by the larvae, which suck juices from plant tissues. If infestation occurs during jointing, infested stems often will break prior to maturity. The Hessian fly is 3-4 mm long, has a black head and thorax, and a pinkish or yellow-brown abdomen.

Life Cycle: Adult flies emerge in the spring from pupae that have overwintered in straw or stubble. The minute, oblong eggs are reddish in color and are laid in rows on the upper sides of leaves. The eggs hatch within one week; the white, legless larvae settle behind the leaf sheaths and suck the sap of the plant. They develop into translucent, pale green, slug-like maggots. The reddish brown pupae, commonly called "flag seed" because of their resemblance to the seed of the flax plant, are oval shaped, flattened, taper to a point, and are 3-5 mm long. They are found behind leaf sheaths, usually at a node.

Hosts/Distribution: The Hessian fly is mainly a pest of wheat, but it may attack barley, rye, and other grasses. This pest has been reported in most wheat-growing areas of the world.

Importance: This is one of the most destructive insect pests on cereals. Widespread outbreaks have occurred and, in some locations (such as North Africa and the USA), the past recurs annually.

Wheat Stem Maggot

Meromyze americana

Symptoms: When young tillers are attacked in the fall or early spring, the tillers usually die; infested plants show the "white head" condition typically produced by stem-boring insects. The adult flies are about 6 mm in length, and pale green to yellow with dark stripes.

Life Cycle: Wheat stem maggot larvae overwinter in cereal plants or grasses. The females lay small white eggs, one per stem, near the sheath of the flag leaf; the larvae burrow into and consume the interior of the stem, killing the upper part of the stem and the head. There are normally three generations per year; one in the spring, one in the summer, and a third in the early autumn that overwinter as larvae.

Hosts/Distribution: In addition to wheat, host crops include rye, barley, and other grasses. There are a number of other flies in various parts of the world that attack wheat in a similar fashion and produce the same kind of damage.

Importance: In infested fields, 10-15% of plants may be injured. Damage can be severe in some years, but the insect seldom causes widespread damage. However, heavy infestations of individual wheat stands may kill a significant portion of the tillers.

Sawfly

Symptoms: Damage by sawflies includes premature

yellowing of the head and shrivelling of the grain. The larvae girdle the stem and, later in the crop cycle, lodging is common.

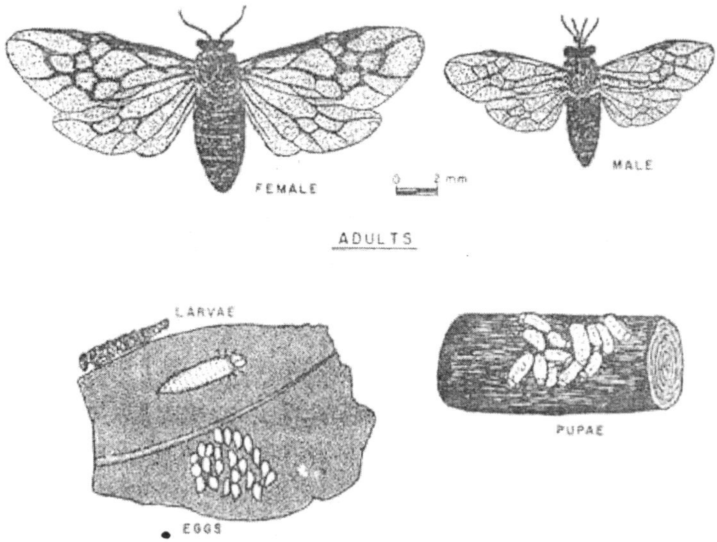

Sawfly

Life Cycle: Sawflies produce one generation per year. The larvae overwinter in the straw; in the spring they pupate. Adult sawflies are small, fly-like wasps and appear from late spring to midsummer. The females deposit small white eggs in the upper nodes of stems just below the heads. Upon hatching, the legless white larvae bore into the stem and tunnel downward, feeding on the pith of the stem. When they have completed their feeding, they descend further and girdle the stem base.

Hosts/Distribution: Nearly all cultivated cereals and native grasses act as hosts, although wheat is preferred. Fall-sown cereals are more commonly attacked. Wheat lines having solid or partially solid stems are much less susceptible to attack. The wheat stem sawfly is a major problem in the Mediterranean Basin.

Importance: Sawfly can cause significant damage in some years, but infestations are usually discontinuous.

White Grubs

Symptoms: White grubs can partially or completely sever the roots of the host plants. This causes patches of wilting and dying wheat plants (especially at the seedling stage), symptoms that could be attributed to root rots. However, when stunted patches are observed, the surrounding soil should be examined for the larvae. When fully grown, the largest of these larvae may be several centimeters long and nearly one centimeter thick. The larvae have three pairs of legs on their thorax.

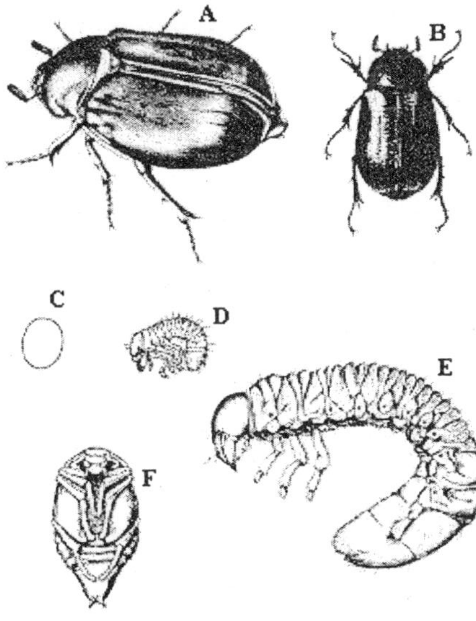

White grubs. A&B, Adults. C, Egg. D,E, Grubs. F, Pupa

Life Cycle: White grubs are the larvae of May or June beetles. Eggs are deposited in the soil and the hatched larvae feed on roots. The duration of the larval stage varies from species to species.

Hosts/Distribution: Many species of white grubs found throughout the world can attack wheat and many other plant species. Cereal crops may suffer significant damage if seeded into heavily infested grasslands.

Importance: When the roots are not completely destroyed, the plants may survive, but are stunted and fail to produce heads. However, the distribution and extent of attack is not uniform.

Wireworms

Symptoms: Wireworm damage is very similar to that caused by other soil-inhabiting chewing insects; the only sure means of identifying wireworms as the causal agent is to find them in association with the damaged seedlings. The name "wireworm" refers to the tough, wire-like appearance of the larvae. They are 20-30 mm long and are often smooth, hard, and highly polished.

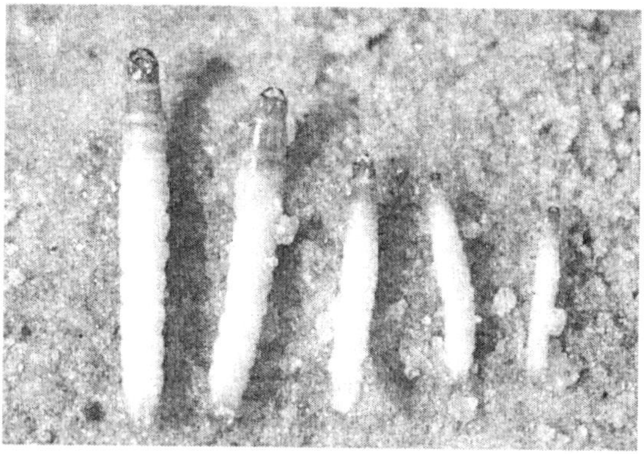

Wireworms

They have three pairs of legs, and their color may vary from a rich cream to shades of brown. Wireworm larvae may attack wheat as soon as the crop is seeded, eating the endosperm of the kernels and leaving only the seed coat. A common sign of woreworm attack is the wilting and/or dying of a number of adjacent plants, either in a row or patch. The stems of affected seedlings will be chewed just above the seed.

Life Cycle: Wireworms are the larvae of click beetles, of which there are many species. The adult beetles lay eggs in the soil, usually in the spring, and the larvae may take several years to develop prior to pupating, depending on species. Generations overlap so that all stages and sizes of larvae may be found in the soil at the same time.

Hosts/Distribution: Many species of wireworms are found throughout the world, all of which can attack wheat. These larvae are capable of attacking many different plant species as well.

Importance: Woreworms are among the most damaging soil-infesting insects. Damage is usually most severe where wheat has been seeded after fallow or after a number of years of grass.

Slugs, Snails, Grasshoppers, and Crickets

Symptoms: Slugs and snails can feed on the endosperm of germinating seed, bite seedlings off at ground level, and graze older plants, chewing longitudinal stripes on the leaves. This gives the adult plant a frayed appearance. Grasshoppers and crickets cause damage that is very similar to that caused by cutworms and armyworms.

Hosts/Distribution: All these insects will attack a large range of plant species, including the cereals. Distribution is worldwide.

Importance: These insects usually are localised in their distribution, but can cause a great deal of damage to individual stands of wheat.

Mites

Symptoms: Adult mites are usually less than 1 mm long, and most of the plant-inhabiting species have sucking mouth parts. Mites are not insects. Adults typically possess four pairs of legs, while larvae have only three pairs. However, the wheat curl mite, *Eriophye tulipae* (syn. *Aceria tulipae*), has only two pairs of legs. When present in large numbers, mites cause a silvery flecking on leaves. Some species may produce webs and/or may cause infested plants to be severely stunted, to head poorly, and to turn white. Individual mites are so small they they can scarcely be seen with the unaided eye. Even so, if an infested leaf is held over a piece of white paper (folded to form a trough) and tapped sharply several time, hundreds of mites will fall onto the paper and can be seen moving about.

Hosts/Distribution: Mites generally are not an important problem, with the exception of the wheat curl mite, which is a vector of wheat streak mosaic virus (WSMV).

BIOLOGICAL CONTOL AND INTEGRATED PEST MANAGEMENT IN WHEAT

Biological control is most effective when used with other compatible pest control practices in an integrated pest management (IPM) program. These practices include cultural control, planting pest resistant varieties and using selective insecticides when other practices fail to keep pest numbers below the economic threshold. These practices and economic thresholds are detailed in the Extension publication *"Managing Insect and Mite Pests of Texas Small Grains,"* B1251, available from your county Extension office.

Biological Agents

Natural enemies of insect and mite pests are classified as parasites, predators or pathogens. Parasites are insects that require only one host to complete their development. The adult typically lays her egg in or on the host. The parasite larva feeds on the host, eventually killing it. The adult parasite then emerges from the dead host's body. Parasites often attack only one or a few related pest species. Many important parasites are small wasps and flies.

Predators are the lions and wolves of the insect world. During their life cycles they eat many insect pests. Predators usually feed on a greater variety of insect species than do parasites. Lady beetles and lacewing larvae are common insect predators.

Insects are subject to diseases caused by viruses, bacteria, nematodes and fungal pathogens. Disease outbreaks can rapidly reduce the populations of pest species when conditions favor infection and transmission. Some pathogens are commercially cultured and sold, and can be used much like an insecticide for pest control.

Methods of Biological Control

The three approaches to biological control of insect and mite pests are importation, conservation and augmentation of natural enemies.

Importation is the release of natural enemies into areas where they do not occur naturally. Some pest species have been accidentally brought into the U.S. from other countries. Natural enemies from the pest's native home are then identified, imported, and released. If successful, the natural enemy will establish permanent populations and control the pest without the need for further releases.

Conservation is the protection and encouragement of existing populations of natural enemies. It is important to recognise natural enemies in the field and be aware of their beneficial actions. Natural enemies can be protected by minimising insecticide treatments and, when possible, selecting products less toxic to beneficial insects. It is sometimes possible to modify cultural practices so that they favor the development of natural pest enemies.

Augmentation is the purchase and periodic release of natural enemies that do not naturally occur in sufficient numbers to provide pest control. As an example, parasitic wasps and predatory mites are commercially reared and sold for the control of pests in greenhouses.

Augmentative releases may be designed to "seed" natural enemy populations by releasing a few insects in the hope that they will increase rapidly to effective numbers (inoculative releases). Or, they may involve the release of large numbers of natural enemies to rapidly reduce pest infestations (inundative releases).

Augmentative releases are often used much like an insecticide. However, because natural enemies are living organisms, their effects are much more complex and variable than those of chemical insecticides. Release programs may fail if too few beneficial insects are released, weather conditions are unfavorable for their survival or reproduction, releases are not properly timed, the beneficials leave the field, or the species released is not adapted to the particular pest or crop situation.

Important questions to ask when considering an augmentation program include:

1. Has research shown that a release program is effective for the particular pest, crop and local situation?
2. Are the proposed release rates sufficient to protect crop yields? When is the best time to release the natural enemy in relation to the pest's life cycle?
3. Will the species of natural enemy to be used attack the target pest?
4. Are releases compatible with the need to apply insecticides for other crop pests and with other crop production practices?
5. Are the natural enemies alive and active when released? What quality control practices does the company use?

6. What directions and assistance does the company provide regarding the handling, release and evaluation of the natural enemy?
7. Discuss your pest situation with several different companies to determine which seems most knowledgeable about its product and your pest problem.

Biological Control in Small Grains

Conservation of existing natural enemies is the most important approach to biological control in small grains. It is important to recognise beneficial insects and consider their impact on pest infestations when making pest control decisions. Using insecticides only when necessary based on field scouting and economic thresholds helps conserve natural enemy populations.

Exotic natural enemies of greenbugs and the Russian wheat aphid have been imported. Texas A&M University is releasing them to establish permanent populations. There is little information on the effectiveness of augmenting natural enemies for control of insect pests in wheat. Until definitive information is available(what natural enemy is effective, release rate, timing, etc.), the Texas Agricultural Extension Service cannot provide guidelines for the use of augmentation for insect control in wheat.

NATURAL ENEMIES OF WHEAT PESTS

Greenbugs

Greenbugs, like other aphids, are attacked by a variety of predators and parasites. These include the parasitic wasps, several species of lady beetles, lacewings and damsel bugs. These beneficials help keep greenbug populations from increasing to damaging levels in many years.

Parasitic wasps: *Lysiphlebus testaceipes, Diaeretiella rapae* and *Aphelinus varipes* are tiny, black wasps which parasitise greenbugs wherever this pest is found. Although the adult wasps are not commonly seen, wheat producers should be able to recognise the distinctive parasitised greenbug mummy which remains attached to the wheat leaf. Greenbug mummies parasitised by *L. testaceipes* and *D. rapae* are beige or tan and are round and swollen. Greenbug mummies killed by *A. varipes* are black and similar in size and shape to live greenbugs.

Lysiphlebus testaceipes is the most common parasite of greenbug in wheat and in some areas is the greenbug's most important natural enemy. Each female adult *Lysiphlebus* can parasitise about 100 greenbugs during her 4 to 5 days of life. More important than the death of individual greenbugs is the reduction in potential greenbug reproduction. All greenbugs are females which give live birth to three or four offspring per day. Parasitised greenbugs stop reproducing within 1 to 5 days, while unparasitised greenbugs reproduce for 25 to 30 days. As a result, parasitism can greatly reduce the rate at which greenbug infestations increase.

Temperature plays an important role in the effectiveness of parasites because parasites develop much more slowly below 65 degrees F and adults are not active below 56 degrees F. Greenbugs remain active at cool temperatures and continue to reproduce until the temperature drops to 40 degrees F. As a result, greenbug infestation can increase unchecked by parasites during cool weather.

Parasite activity in the field can be monitored by looking for greenbug mummies on wheat leaves. Weather conditions will largely determine how quickly parasites can prevent a greenbug outbreak. Remember that aphids that appear healthy may actually have parasites developing within, as the mummy stage does not develop until 8 to 10 days after parasitism. As a general rule, a greenbug infestation declines rapidly once 20 percent of the greenbugs are mummies, because at this point most of the living greenbugs are already parasitized though they have not yet entered the mummy stage.

Insecticides applied as sprays will kill adult wasps and indirectly kill immature parasites by killing their greenbug hosts. Methyl parathion and chlorpyrifos (Lorsban®) are more toxic to adult wasps and to immature parasites inside greenbugs than the systemic insecticides dimethoate and disulfoton (Di-Syston®), especially at lower rates. However, the short residual toxicity of methyl parathion allows parasites to recolonize a field sooner than when insecticides with longer residual activity are used.

Immature parasites inside greenbug mummies gain some protection from insecticides. Parathion is more destructive to parasites in mummies than dimethoate or disulfoton (Di-Syston®). Chlorpyrifos (Lorsban®) is reported to be highly toxic to *Lysiphlebus* inside aphid mummies. Higher insecticide rates increase the mortality of immature *Lysiphlebus* and extend the time before adults can survive on treated foliage. The fungicide Bayleton®, used to control leaf rust on wheat, also is very toxic to adult parasites.

Parasites

Lysiphlebus testaceipes

A shiny black wasp slightly smaller than a mature greenbug. The wasp can be seen on warm, sunny days crawling across wheat leaves and stinging (parasitising) greenbugs. Parasitised greenbugs are swollen and tan and are called "mummies."

Role: One of the most important natural enemies of the greenbug in wheat and sorghum. Once about 20 percent of the greenbugs are in the mummy stage, the infestation usually declines rapidly. This parasite is most effective during warm weather in the fall and early spring; adults are inactive at temperatures below 56 degrees F. The parasite also attacks the corn leaf aphid, cotton aphid and others.

Life cycle: The female pierces (stings) the greenbug and deposits an egg inside. In about 2 days the tiny egg hatches into a grub which feeds internally on the living greenbug. The parasite becomes mature in about 6 to 8 days and then begins to twist and turn inside the greenbug host. The parasitised greenbug "seizes the leaf with a rigid death-grip and the last spark of life soon fades." Movement of the parasite larva expands and molds the greenbug's body, giving it a swollen appearance. The parasite larva cuts a hole through the bottom of the greenbug's body and fastens it to the leaf surface with silk and a glue. The swollen greenbug cadaver turns a beige to tan color and is then called a "mummy."

The parasite larva then enters the pupa stage inside the mummy and the adult parasite emerges 4 to 5 days later. The adult escapes through a circular hole it cuts in the top of the mummy. The adult parasite mates and repeats the life cycle by stinging additional greenbugs. At a constant temperature of 70 degrees F, development from egg to adult requires about 14 days.

The parasite overwinters as a larva or pupa inside a parasitised greenbug. Wasps disperse by flying and by being carried as immature parasites inside winged greenbugs. Unfortunately, *Lysiphlebus* is often attacked by other species of parasitic wasps, which can limit its control of greenbug.

Lady Beetles: Several species of lady beetles can be found in Texas wheat, including the convergent lady beetle (*Hippodamia convergens*), the seven-spotted lady beetle or C-7 (*Coccinella septempunctata*) and the much smaller species *Scymnus loewii.*

As a group, lady beetles provide some control of aphids in small grains every season. But since lady beetles consume the aphids they leave no evidence of their control. Sometimes the only indication is the lack of significant aphid problems later in the year. No one lady beetle species can completely reduce an aphid population, and research has shown that augmentation by releasing lady beetles is of little benefit. The most significant action a farmer can take to help lady beetles control aphids is conservation. Conservation primarily means avoiding broad-spectrum insecticides and using the lowest rate of insecticide that achieves adequate control of the target pest. It is also helpful to apply insecticides when the lady beetle population consists of egg or pupal stages rather than young larvae.

Convergent lady beetles alone or in combination with parasitic wasps often suppress greenbug and bird cherry-oat aphids in wheat in the Texas Rolling Plains. The number of lady beetles that develop in a field depends on environmental conditions and the size of the aphid population. Lady beetles have the greatest effect on greenbugs before temperatures become extremely cold in November and when temperatures warm up after mid-February. During extremely mild winters, lady beetles may be active in wheat fields in December or January. Small greenbug populations have been suppressed by lady beetle densities of one beetle per 10 feet of drill row.

In California, huge numbers of convergent lady beetles hibernate in masses on mountain sides and are easily collected and stored until sold in the spring. As early as 1910 farmers were purchasing lady beetles from California and releasing them in an attempt to control aphids. In their native environment, however, beetles emerging from hibernation fly from the mountains to return to summer habitats at lower elevations in search of food. Thus, when released into fields after storage, the beetles naturally respond by taking flight. Studies have shown that 99 percent of released lady beetle adults fly from the release site; attempts to retain them by providing shelter and water and by releasing at night have been unsuccessful.

The C-7 lady beetle is larger than the convergent and is bright red with seven black spots. This species was introduced into the U.S. from Europe and is now found throughout Texas except for far west and southwest Texas. Adult females live for about 4 to 6 weeks and produce about 1,000 eggs. C-7 larvae feed for about 2 weeks, consuming about 20 greenbugs per day in laboratory studies.

Scymnus is a small, dull brown beetle with a V-shaped black area down the centers of the wing covers. Larvae are covered with long, waxy filaments and feed on aphids and mites.

Heavy, rapidly increasing greenbug infestations can cause excessive damage; however, lady beetles and parasitic wasps, under favorable weather conditions, can reduce greenbug populations to below treatment levels. Where there are one to two lady beetles (adults and larvae) per foot of row or 15 to 20 percent of the greenbugs have been parasitised, control measures should be delayed until it can be determined whether the greenbug population is continuing to increase. When weather conditions permit these beneficials to remain active, they will significantly reduce the greenbug population during the following week.

Convergent lady beetle, Hippodamia convergens

This is the common orange lady beetle with six small black dots on the wing covers. The black plate (pronotum) just behind the head has two diagonal white lines which appear to converge, thus the common name convergent lady beetle. The bright yellow eggs are laid in groups on leaves. Larvae are alligator-shaped and black with rows of orange spots.

Role: Convergent lady beetle adults and larvae feed primarily on aphids but also eat other insect eggs and larvae. Under laboratory conditions, convergent lady beetle larvae consume about 20 greenbugs. Like the parasitic wasps, lady beetles are most effective while temperatures are warm in the fall (November) and again after mid-February. Lady beetle densities of one per 10 feet of drill row can suppress small greenbug infestations during mild weather in the Texas Rolling Plains. This species of lady beetle is less effective on Russian wheat aphids because they are too large to reach Russian wheat aphids which are protected inside rolled wheat leaves.

Life cycle: Development from egg to adult requires 3 to 4 weeks, depending upon the temperature. Eggs hatch in 3 to 4 days. Larvae feed for about 2 weeks and then enter the pupal stage, which is immobile and attached to a leaf. The adult beetle emerges from the pupa in about a week.

Lacewings: Green lacewings are found in many crops, including wheat. However, there is little information on the importance of lacewings in the biological control of greenbugs. The larval stages of all lacewings are predators which feed primarily on aphids but also on insect eggs and small caterpillars. The adults of some lacewing species, such as *Chrysoperla carnea,* feed only on honeydew or pollen. Lacewing larvae are not common in wheat but compliment other predators. The number of lacewing adults usually peaks in April and May in wheat fields. The brown lacewing also occurs in wheat.

Green lacewing fly, Chrysoperla carnea

Adult green lacewings are delicate, slender insects 1/2 to 3/4 inch long with large wings held roof-like over their bodies. The green wings have a fine, net-like pattern of veins. Their eyes are a bright gold color.

Larvae are alligator-shaped, grayish-brown and have long sickle-shaped mandibles protruding from their heads.

Newly hatched larvae are very small (less than 1/8 inch long), while mature larvae are 1/4 to 3/8 inch long.

Eggs are laid singly on top of a fine thread attached to leaves or stems and are easily seen and recognised.

Role: Lacewing larvae feed primarily on aphids in wheat but also attack small caterpillars, insect eggs and mites. They are most common in wheat in the spring. Although the adults of some species are predators, adult *Chrysoperla carnea* feed only on honeydew, nectar and pollen.

Life cycle: Eggs hatch in 3 to 6 days. Larvae are voracious predators and will eat other lacewing eggs and larvae. Larvae feed for about 2 weeks and then spin a spherical white cocoons of tough silk. The larvae pupates inside the cocoon and the adult lacewing emerges in about 10 days.

Upon emergence, adult lacewings make long dispersal flights regardless of the presence of food. Lacewings fly at night and may travel several miles. Females must feed on nectar, pollen or aphid honeydew to produce eggs. Females lay their first eggs 4 to 6 days after emergence. Adults live several weeks, during which they lay 200 or more eggs.

Nabid or Damsel Bug

Nabids are slender, dull tan to grey insects about 5/16 inchlong. They have long legs and antennae.

Role: Nabids feed on greenbugs, other aphids and soft bodied insects. They are found in wheat throughout the winter but only as the adult stage. Adults do not begin to reproduce until about April and, as a result, densities of nabids cannot increase earlier in the season in response to an increasing greenbug infestation.

Russian wheat aphid

The Russian wheat aphid is a new pest in the U.S., having been first reported in 1986 near Muleshoe, Texas. Many of the natural enemies of greenbug also attack Russian wheat aphid. However, lady beetles such as *Coccinella* and *Hippodamia* are too large to enter the rolled leaves in which Russian wheat aphids are found. There may be natural enemies in this pest's original range that could be imported for biological control. Entomologists began searching for natural enemies of Russian wheat aphid in 1988 in the Middle East and central Asia, where this pest is believed to have originated, and in Europe. Since then, several promising natural enemies have been collected, reared and released in Texas and elsewhere. A species of parasite imported from France, *Aphelinus asychis,* has been recovered near Amarillo. *A. asychis* is a very small parasite that can enter the curled leaves of wheat plants where Russian wheat aphids feed. Researchers are working to determine how well it and other released natural enemies may be controlling Russian wheat aphids.

As with greenbugs, predators and parasites often provide enough control to keep Russian wheat aphids below the economic injury level. Therefore, it is important to conserve natural enemies. Before an insecticide is applied, the farmer should properly assess the density of an aphid population and determine whether it is increasing or decreasing. If the aphids are increasing rapidly and few lady beetles and parasites are present, an insecticide application may be necessary.

Insect predators and parasites are very small and are often effective at low numbers. When they are effective, pest numbers do not become threatening and, as a result, the benefit of these natural enemies often goes unnoticed. The importance of natural en emies can be experimentally measured by excluding them with cages. The graph below shows the number of Russian wheat aphids in wheat when natural enemies were present or absent. Parasitic wasps, lady beetles and other predators could not enter the closed cages and, as a result, the number of Russian wheat aphids rapidly increased in late May. In contrast, the open cages allowed natural enemies to enter and attack Russian wheat aphids. As a result, aphid numbers were much lower.

REFERENCES

Belderok, Bob & Hans Mesdag & Dingena A. Donner, *Bread-Making Quality of Wheat*, Springer, 2000.

Smith, Albert E., *Handbook of Weed Management Systems*, Marcel Dekker, 1995.

Smith, C. Wayne, *Crop Production*, John Wiley and Sons, 1995.

3

Insect Pests of Sorghum and their Management

Grain sorghum is relatively tolerant to insect feeding. Higher seeding rates and ability to tiller make stand losses to soils insect less severe than in corn. Likewise, sorghum is very tolerant of defoliation. However, insects that prevent seed set or cause direct damage to seed can produce serious loss, because the crop does not have time to compensate by producing new florets and setting seed. Treatment thresholds for most insect pests usually are high. The crop's tolerance of injury with little yield loss coupled with low commodity prices usually do not justify the cost of the application of an insecticide. Therefore, if a "borderline" economic infestation exists of any pest or pest complex – do not treat.

GENERAL MANAGEMENT

Rotate crops. Rotation will help minimise the buildup of sorghum pests in the same field.

Select proper hybrids. Plant well-adapted, vigorous, high-yielding hybrids with good disease resistance and standability.

Maintain fertility. Test soil and apply fertiliser and lime according to the University of Georgia Cooperative Extension Service recommendations to maintain the proper soil fertility. Samples can be taken to your local county extension office. A minimum charge is required for testing and recommendations.

Plant early (when possible). Sorghum planted in early May in south Georgia and by the middle of May in north Georgia often escapes major insect damage. Several sorghum pests, such as sorghum midge, do not usually reach damaging population levels until after early plantings are mature.

Make uniform plantings of uniformly maturing hybrids over as large an area as possible. By having large acreages of sorghum at about the same developmental stage, infestations of these insects are "diluted" rather than being concentrated in one field or in a small part of the acreage which may be more attractive to pests. Small irregular fields also are at greater risk of bird damage than larger fields.

In conventional-tillage, bury previous-crop residue (especially sod) and keep free of weeds (especially grasses) for at least 2 weeks before planting. This practice helps reduce the incidence of damage from insects, which may establish infestations on weeds, volunteer grasses or grass sod. Also, cutworms are often a problem where previous-crop residue is present on the soil surface at planting.

Harvest timely in order to prevent seed damage by pests and birds.

INSECT PESTS

Important pests of grain sorghum in the southern United States include the sorghum midge, lesser cornstalk borer, soil insects, fall armyworm, headworms, chinch bugs and stink bugs.

Soil and Seedling Pests

Soil insects

Soil insects that attack sorghum include wireworms, white

grubs, whitefringed beetle larvae, seedcorn maggots, southern corn rootworm, and lesser cornstalk borer. Soil insects are difficult to sample and predict. Before planting, inspect soil carefully when preparing land for planting (especially when turning land) for the presence of white grubs, wireworm or whitefringed beetle larvae. Since these insects will go back to the soil within seconds, make inspections immediately after turning the soil. Inspect the root systems, and soil around the roots, of any weeds present for these insects and their feeding damage. Soil insects will destroy seed causing gaps in the stand. They also can bore into seedling plants below the soil surface killing the main stem, which kills the leaf whorl causing a "deadheart" type injury.

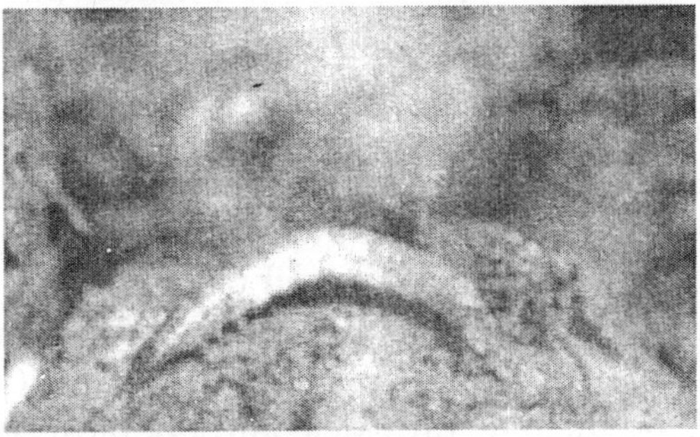

Wireworm.

The following conditions increase the risk of damage by soil insects:

1. Planting late such as doublecropping behind winter crops or vegetables.
2. Planting into sod or following newly tilled ground previously in sod or weedy species.
3. Doublecropping behind cereal grains or winter cover crops. Cereal crops and sorghum are both grasses and are often attacked by the same insects.

4. Reduced, strip-till and no-till systems where previous crop residue remains on the soil surface. Cool wet conditions favor wireworms, grubs and cutworms.
5. Planting into burnt cereal stubble and/or planting under conventional tillage on light soils during hot, droughty conditions. This favors lesser cornstalk borer infestations.

Sorghum can tolerate some stand loss without affecting yield. The higher seeding rates allow sorghum to withstand up to 15 percent stand loss. However, soil insect damage often is not uniform and is concentrated in certain areas. Losses in those spots can be severe.

Soil insecticides are not routinely used at planting in grain sorghum. However, if one or more of the above conditions exist, it is appropriate to consider the use of an at-planting insecticide or seed treatment. The best way to apply an at-planting granular insecticide on grain sorghum (where the label allows) is in a narrow band over the row behind the planter shoe and in front of the press wheel. This method of application concentrates some of the material in a band near the surface of the soil where insects are more likely to come in contact with it. New systemic insecticide seed treatments, such as Gaucho™ and Cruiser™, depending on the rate, provide control of most soil insects while also systemically protecting the seedling for the first few weeks (20-25 days) after planting. Seed treatments usually must be applied to seed by specialised seed treatment equipment but do not require specialised application equipment on planters. They also have better worker safety than older organophosate insecticides.

Seedling Pests

Several different insects can attack, damage and kill seedlings. Insect feeding will kill the plant or destroy the primary tiller causing the plant to produce secondary tillers or suckers. These pests include:

Lesser cornstalk borer (*Elasmoplapus lignosellus*) is a moth. Larvae feed below the soil line where they bore into the stem base often killing the main growing point. Larvae are blue-green with transverse brown bands. LSCB is easily identified because larvae are usually enclosed in a silken tube, which has incorporated soil particles. LCSB is favored by conventional tillage under hot, dry conditions, sandy soils, double-cropping behind cereal grains, and planting into burnt cereal grain stubble. Preventive insecticides best control LCSB. Rescue treatments, once damage is detected, often are ineffective.

Billbugs (*Sphenophorus* spp.) are medium brown to blackish beetles with a distinct beak typical of weevils. Adults chew holes in the stem base usually at of near the soil line. Infestations often are associated with nut sedge, which is the only significant alternate host. Billbugs normally crawl from field to field, and infestations often occur annually in the same field. Consequently, crop rotation can be effective at controlling this pest.

Sugarcane beetle (Euetheola humilis rugiceps) is a small black scarab beetle that chews a hole in the stem usually at or near ground level. This beetle can attack seedling and older stage plants. It is often associated with infestation of bahiagrass, which is a preferred alternate host.

Sugarcane beetle

Cutworms (several species) are larvae of moths. Several species may attack sorghum. Typically they remain in the soil and feed at night where they cut off seedling plants at or near ground level. Plants cut above the main growing point may regrow. Heavy infestations usually kill the main stem or completely destroy plants. Risk of cutworm damage is greater in reduced tillage systems. Control weeds several weeks before planting. Cutworms can be controlled preventively using at-planting soil insecticides or by at-planting or post-emergence foliar sprays. Spay in a band over the row.

Cutworm

When to treat: Treat when damage exceeds 15 percent from seedling emergence until plants are 12 inches tall or when damage is expected due to the planting situation. In addition to soil insecticides and systemic seed treatments, foliar insecticides may control billbug and sugarcane beetle. Spray should be directed at the base of plants with at least 20 gallons per acre of finish spray.

Chinch bug (*Blissus leucopterus leucopterus*) can infest sorghum at any stage, but is most damaging during the seedling and early vegetative stages. Adults are about ?-

inch long, black with a white to silver X marking on the back. Nymphs are reddish brown when small becoming dark gray just before adulthood. Nymphs also have a white stripe across the back.

Chinch bugs suck plant sap and feed mostly on the stem and large leaf veins. Their feeding causes red/ yellow leaf discoloration, wilting and stunted growth. Large numbers can kill seedlings. During grain filling large infestations can kill leaves hastening plant senescence. Chinch bugs move from nearby cereal grain fields to invade sorghum fields in spring and early summer. In this situation chinch bugs usually will concentrate along field margins or in one area of a field. They also may reside on grassy weeds in reduced tillage and move to seedling plants as they emerge. Chinch bugs prefer hot dry conditions. Cool, wet conditions and heavy rain usually promote a fungal pathogen that can cause substantial mortality and therefore chinch bugs are not a problem under these conditions.

When to treat: In seedlings, treat when two or more adults are found on 20 percent of seedlings. On taller plants up to 6 inches, treat when 75 percent of plants are infested OR five or more chinch bugs per plant are present. Chinch bugs usually reside at the base of plants and under the leaf sheath. Consequently, post-emergence applications should be directed at the base of plants using at least 20 gallons per acre of spray. Getting good spray coverage becomes more difficult in larger plants. Treatments after the boot stage rarely are effective. Remember, at-planting applications and systemic seed treatments may provide seedling control for 20-25 days.

Leaf and Stalk-Boring Pests

Aphids and greenbug

Several species of aphids can attack sorghum. Aphids are small soft-bodied insect that have sucking mouthparts and feed on plant sap. Predominant aphids attacking sorghum in Georgia are the corn leaf aphid, *Rhopalosiphum maidis*, and the greenbug, *Shizaphis graminim*. Yellow sugarcane aphid, *Sipha flava*, also occasionally is a pest in Georgia, but is more prevalent in southern Alabama. Corn leaf aphids are blue greenish with black legs and antennae. Corn leaf aphid feeding can reduce plant growth by removing sap but is not toxic to the plants. Consequently sorghum can support large corn leaf aphid infestations without much damage. Greenbug are medium green with a distinct dark green stripe down the back. Yellow sugarcane aphids, are yellow with small black dots. Feeding by greenbugs and yellow sugarcane aphids causes a toxic reaction in the plant causing brown lesions. Large infestations can stunt and kill leaves and whole plants.

Aphids also secrete a sugary solution called *honeydew*. A black fungus called sooty mold can grow on the honeydew. The fungus does not directly harm the plant but blocks sunlight from contaminated leaves and can discolor developing seeds.

When to treat: Aphids seldom require control in Georgia. Usually natural enemies such as lady beetles, hover fly larvae, parasitic wasps and others will control aphid infestations. However if aphids, especially greenbug and yellow sugarcane aphid, persist and are discoloring and stunting seedlings, an insecticide treatment may be needed. In larger plants, an insecticide treatment may be needed if aphids are causing the discoloration and death of two or more leaves. Systemic seed treatments normally are not justified specifically for aphid control, but if used they will control aphids for about 20 days after planting.

Whorl-worms, Budworms, Fall Armyworm in Whorls

Fall armyworm (*Spodoptera frugiperda*) is the most common caterpillar feeding in the whorls of sorghum. True armyworm (*Pseudaletia unipuncta*), corn earworm

(*Helicoverpa zea*) and other armyworms (*Spodoptera* spp.) also may infest whorls. Fall armyworm moths lay egg masses on the plants. Small larvae will feed on leaves causing windowpane type feeding before moving into the leaf whorl. There they will feed until full size when they leave the whorl and pupate in the soil. Larvae are variable in color, usually are dark with a cream to white inverted Y on its head, three white lines along the body, and 4 dots in a nearly square pattern on the top of the last abdominal segment. Larvae can reach 1.5 inches long.

Fall armyworms

When fall armyworms feed in the whorl they are called budworms or whorl-worms. Feeding can cause large irregular holes in leaves as they unroll from the whorl. Sometimes the main vein is cut causing the entire leaf tip to die. Fall armyworm populations increase as the season progresses. This is why late-planted or double-cropped sorghum is at greater risk of defoliation than early-planted sorghum.

When to treat: Sorghum is very tolerant of defoliation. Insecticide control seldom is justified. Consider treatment when 75 percent or more of the whorls are being damaged and larvae are present, OR one or more worms per whorl are present. To obtain effective control, spray applications should be applied by ground equipment using cone nozzles to direct 25-30 gallons of spray into the whorls.

Grasshoppers

Nymphs and adults of several species of grasshopper may chew holes in leaves causing a ragged appearance in sorghum. Grasshoppers can buildup in field border areas and move into sorghum fields, consequently damage often is concentrated along field margins. Grasshopper populations usually are greater in dry than wet years.

When to treat: Consider an insecticide treatment when eight nymphs per square yard are present and damage is evident. Sometimes only a border treatment of the first 50-100 feet may be needed if grasshoppers are concentrated in the field margin.

Mites

Spider mites (usually *Tetranychus urticae*) occasionally will damage sorghum in the Southeast. Mites are very small arachnids with adults having eight legs and larvae having six legs. Spider mites feed on the lower leaf surface where they suck plant fluids. They inhabit webbing usually on the lower leaf surface. Feeding causes leaf wilting and chlorosis (discoloration) as their feeding destroys plant cells. Mites prefer hot, dry conditions.

When to treat: Mites seldom require control in Georgia. Treat if infestations are present and leaf discoloration is apparent. Chemical treatments after the boot stage are rarely effective and not recommended.

EUROPEAN CORN BORER AND OTHER STALK BORERS

Several stalk boring moth caterpillars that attack corn, including European corn borer, *Ostrinia nubilalis;* southwestern corn borer, *Diatraea grandiosella* and others, also may infest sorghum. In areas where sugarcane is grown such as the southernmost counties of Alabama, sugarcane borer, *Diatraea saccharalis*, may attack nearby sorghum fields.

These moths lay eggs in masses on plants. Small larvae feeding on leaves cause shot-hole defoliation before tunneling into the stalk. Larger larvae tunnel in stalks causing reduced grain fill and increased stalk lodging. Stalk borer damage also may make plants more susceptible to stalk rot diseases.

European corn borer has two distinct generations in Georgia. First-generation larvae bore into whorl leaves and sheaths, while second-generation larvae bore into the main stalk. Later plantings are at greatest risk of damage. These moths overwinter as large larvae in crop stubble. Disking or shredding stalks will help reduce winter survival and populations the following year.

When to treat: European corn borer and other stalk borers rarely cause damage to sorghum in Georgia. To be effective, insecticides must be applied to kill larvae before they tunnel into stalks. Treatment may be warranted if egg masses are numerous and/or newly hatched larvae are causing shot hole damage on 25 percent or more of plants. However, foliar applied treatments are only at best 50 percent effective. Because of the cost of sampling, infrequent infestations and difficulty of timing treatments, chemical control is rarely justified.

Panicle and Seed Pests

Sorghum midge

Sorghum midge (*Stenodiplosis sorghicola*) has the potential to be a devastating pest of sorghum in Georgia. The adult is a tiny, fragile-looking, orange-colored gnat-like fly. Females lay eggs singly in sorghum flowers during pollination.The larva is an orange maggot that feeds on the developing seed. When larvae are full size they pupate within the flower. The pupa pushes its way to out and adults emerge leaving the pupal case protruding from glumes. The florets have a "blasted" appearance. The entire life cycle can take as few as 14 days.

Sorghum and johnsongrass are the only hosts of sorghum midge. Sorghum midge populations can build up on johnsongrass before sorghum planting. Sorghum midge also will increase throughout the season on successive plantings of sorghum. Consequently, late-planted sorghum is at increased risk of sorghum midge damage.

Control of sorghum midge requires control of egg-laying adults. Timing of control is critical because maturing adults occur in the morning. Sampling and spray(s) need to be applied at that time for effective control. Sample fields when plants reach 20 percent bloom and continue sampling every few days until 90 percent bloom. The most efficient technique is to carefully inspect sorghum panicles (heads) for adults. Sample at least 20 panicles for each 8 acres of field. Panicles also can be shaken into a plastic bag and the contents inspected for midges.

Management tactics for sorghum midge include crop rotation, early planting, controlled seed depth so that plants emerge uniformly and fields flower at the same time, controlling johnsongrass, and plowing to bury overwintering larvae to reduce midge emergence the next season. Plant resistance has been identified that reduces damage by about 80 percent. Resistance, however, is not currently available in commercial hybrids grown in this area.

When to treat: The treatment threshold for sorghum midge is one adult per panicle (head) at about 25-30 percent bloom. If sorghum prices are high a threshold of one adult per two heads should be used. Re-inspect fields within three days and continue to sample until 90 percent bloom. Re-treat as needed if one or more adults per head are found. When the economic threshold is reached, spray immediately; delay of a day or two can result in serious damage. The susceptible period for midge is during bloom; treatment after bloom is too late. Larvae cannot be controlled once eggs are laid in the flower. Most labeled insecticides are effective against sorghum midge at the lowest labeled rate.

Headworms (Corn earworms, Fall armyworms, Sorghum Webworms)

Headworms usually are a mixture of several species of caterpillars. The main species are corn earworm (*H. zea*), fall armyworm (*S. frugiperda*) and sorghum webworm (*Nola sorghiella*). Larvae of these moths feed on flowers and developing seed, where they partly or completely consume seed. In most years, the corn earworm is the most important species partly because its large size allows it to consume a lot of seed. Fall armyworm can also cause considerable damage during outbreak years.

Corn earworms are greenish to brownish with a yellowish head and true legs. Fall armyworms are darker with a distinct inverted white Y on their heads, three white lines along the body, and four dots in a nearly square pattern on the top of the last abdominal segment. Sorghum webworms are smaller than the other two, flattened, yellow-greenish with four lengthwise dark stripes. Mature webworms are covered with hair that can be irritating to some people. Corn earworms and armyworms occur individually on panicles. Sorghum webworms often are clumped on one or a few panicles. Because headworms are exposed on the heads, they are attacked by numerous parasites and predators including birds. Under the right environmental conditions, large infestations of headworms also can be decimated by several insect diseases.

These insects have several generations per year and populations increase throughout the season. Consequently, the risk of head worm damage greatly increases in later plantings. After sorghum begins to bloom, inspect field once or twice a week. Sample by shaking heads over a bucket or sweep net and counting larvae. Sample 30-50 panicles (heads) per field.

When to treat: Treat when a total of two or more larvae (½ inch or longer) of corn earworm and fall armyworm occur per head. Treat if five or more webworms per head are found.

Stink Bugs

Sorghum seed is readily attacked by several species of stink bugs and related insects. Common stink bugs in sorghum include the southern green stink bug (*Nezera viridula*), brown stink bug (several species of *Euschistus* spp.), and rice stink bug (*Oebalus pugnax*). A related insect, the leaffooted bug (*Leptoglossus phyllopus*), also causes the same type of injury. Leaffooted bug is large about 1 inch long, brown with a thin white stripe across the back. Their hind legs are dilated and flattened and resemble a leaf.

Stink bugs usually move into sorghum soon after flowering. They lay barrel-shaped eggs in clusters. Nymphs will develop and typically complete development before sorghum maturity. Stink bugs have sucking mouthparts and feed on developing seed within the panicle. Their feeding will shrivel seed thereby reducing seed size and quality. They also may feed on stems and other plant parts but the main damage is from kernel (seed) feeding. Sorghum is most susceptible during the milk and soft dough stage. Injury normally is not damaging from hard dough to maturity.

Stink bugs can be sampled using the same techniques described for headworms. Sample by beating or shaking panicles over a sweep net or bucket and counting the bugs that appear. Sample 30-50 panicles per field.

When to treat: Treat if combined numbers of all stink bugs (large nymphs and adults) exceeds four bugs per head during milk stage or eight bugs per head during soft dough stage. Insecticides recommended for stink bug control also will control headworms. However, some of the insecticides recommended for headworms such as spinosad (Tracer™) are not effective against stink bugs.

False Chinch Bugs

False chinch bugs are related to true chinch bugs, but are a uniformly medium gray. False chinch bugs can occur in large numbers during grain filling. They suck plant fluids and can injure developing seed. Their small size, ¼ inch, means large numbers are needed to damage sorghum. Scout using the same techniques described for stink bugs and headworms. Infestations usually are concentrated in one area of a field.

When to treat: Treat if populations exceed 140 bugs per head during milk to early soft dough stage. False chinch bugs rarely require control by themselves, but should be considered in combination with stink bugs and headworms.

Birds

Several types of birds including blackbirds can infest grain sorghum during the period of hard dough to maturity, where they perch on panicles and eat seed. Birds will consume whole seed but also will break seed by eating portions and leaving half of the seed on the panicle. This damage is easily recognised because of the clean break and exposed white endosperm of the seed. Birds also will defecate while feeding, and their droppings and feathers are readily apparent on the panicles and upper leaves.

Small fields surrounded by woods and fields with structures such as power lines and irrigation equipment provide birds a place to perch and congregate when not feeding. Late-planted fields with maturing grain in the fall, when birds start swarming, also are at increased risk of bird damage. Birds quickly learn the location of fields when sorghum is planted year after year in the same location.

There are no legal chemical methods for bird control. Bird distress callers, blank shooters and various types of reflective plates and tinsel may scare off birds. Timely harvest will help reduce the potential for bird damage.

BENEFICIAL INSECTS

Several types of beneficial arthropods commonly occur in sorghum. They will help control pest species. Correct identification is important so that beneficials are not mistaken as pests. Beneficial insects in grain sorghum generally are the same species as those in other field crops. Common beneficial arthropods include the following: lady beetles, syrphid or hover fly larvae, big-eyed bugs, damsel bugs, minute pirate bug, lacewings, ground beetles, spiders and numerous parasitic wasps and flies.

REFERENCES

FAO Report, "Sorghum and millets in human nutrition", 1995.

Watson, Andrew M., *Agricultural Innovation in the Early Islamic World: The Diffusion of Crops and Farming Techniques,* Cambridge: Cambridge University Press, 1983.

United Nations Food and Agriculture Organization: Sorghum and millets in human nutrition.

4

Barley Pests and their Control

Weeds are responsible for serious yield losses in barley and controlling them will usually result in a substantial increase in net income. Weeds compete with barley for light, nutrients and moisture. When these factors are limiting, barley yields will be reduced. The yield reduction will be directly proportional to the weed population in the crop. In many areas of Alberta, yields are still reduced by 15 to 20 per cent through competition from weeds.

Wild oats constitutes a major weed problem to many barley growers in Alberta. Research on barley in central Alberta shows that yield losses through competition from wild oats may be predicted when the number of wild oat plants present per unit area is known. This is done by applying the equation:

L = 0.023 times A times square root of X

where L is the predicted yield loss (g/m^2)

A is the predicted weed-free yield (g/m^2)

X is the number of wild oat plants per square metre. (5.38 g/m^2 = 53.8 kg/ha = 1 bu/ac)

For example, if a field of barley would yield 322.8 g/m^2 (60 bu/ac) with no wild oats present, but it is found to have 30 wild oat plants per square metre by actual count, then the predicted yield loss can be calculated as follows:

L = 0.023 times A times square root of X

A = 322.8

X = 30

therefore L = 0.023 x 322.8 g/m^2 x sq rt of 30

L = 40.67 g/m^2 (7.6 bu/ac) yield loss.

Weed seeds present in barley grain may be responsible for major dockage and grade losses at market time. Excessive weed growth present in barley harvested for silage may reduce the quality of the fodder and can affect palatability. Malting barley contaminated with excessive numbers of certain weeds will not achieve malting grades. The cost of control must be balanced by the expected increased return to the producer from an increase in crop production, both in the current year and in future years.

METHODS OF CONTROLLING WEEDS

There are five fundamental methods of controlling weeds in barley.

Preventive Measures

- Use weed free seed.
- Use only clean, weed-free machinery and hauling facilities.
- Restrict animal movement from weedy fields to clean fields.
- Handle weedy manure in a manner to eliminate spread of viable weed seeds.
- Where possible control or restrict wind and water movement from weedy to weed-free areas.

Physical Control Measures

- Tillage that will bury weeds, expose them to frost or

drying, or wound them to allow for entry of decay-causing organisms.

— Hand weeding of smaller patches of select weeds.
— Mowing where tillage is impractical.
— Burning weeds prior to maturity.
— Pasturing and grazing to prevent seed formation.
— Mulching small weed patches with clean straw, manure or polyethylene film.

Cultural Control Measures

— Use crop rotations that may include fallow, forage crops, row crops, or fall-sown cereals, to take advantage of weaknesses in the life cycle of existing weeds and thereby eliminate them.
— Plant competition. Any measure that enhances vigorous crop growth helps to control weeds.
— Adjust timing of cultural operations. Delayed seeding has traditionally been used to control certain weeds.

Today, however, with the herbicides available, delayed seeding is a very doubtful economic practice for controlling weeds. Seeding fall-sown crops in rotation with barley is an effective weed control ,measure. Crop growth is established before weed growth begins in the spring.

Biological Control Measures

This technique has not yet been perfected to the point where it has any major value to individual barley producers. Control agents are usually insects or diseases which attack only one specific weed.

Chemical Control Measures

Many herbicides are available for use on weeds commonly found in barley in Alberta. Most herbicides, currently recommended, are relatively cost-efficient and are the only widely applicable means of controlling weeds in the growing crop.

When herbicides are used as directed, they usually do a good job of controlling a wide range of weeds, though some are specific and control only a few species.

Most, though not all, failures to properly control weeds can be explained by the applicator failing to follow the label directions.

From surveys of farmers in agro-climatic zones 1,2 and 3, the four most troublesome weeds were Canada thistle, wild oats, hemp nettle and chickweed. This listing was the same for the top and low producers. The differences noted between weed control measures used by top and low producers were:

1. Top producers achieved greater weed control at the seedling stage and throughout the growing season.
2. Top producers were more aware of their weed problems (types and numbers).
3. The top producers achieved better weed control through greater use of broadleaf herbicides, including mixes, and a greater use of wild oat herbicides.
4. Only 20 per cent of the top group felt that wild oats contributed to yield loss after their herbicide program, compared to 55 per cent in the low group.

BARLEY DISEASES

Major diseases in barley are pathogenic in nature and are caused by fungi, bacteria or viruses. Other diseases are non-pathogenic and are caused by harmful environmental factors such as nutrient deficiencies or adverse weather. Pathogenic diseases are controlled most easily by using resistant varieties, but these are not always available. In some instances, the development of resistant varieties by plant breeders is nearly counterbalanced by development of new races of the pathogen.

Common Root Rot

Common root rot is consistently one of the most damaging diseases of wheat and barley. It can also infect oats, rye and triticale. There are a number of phases to this disease complex; seedling blight, root rot and later in the season, a possible leaf spot phase. The seedling blight phase is less common on wheat than on barley. In the mature plant, root rot caused primarily by *Cochliobolus sativus* is the major cause of crop loss. The leaf spot and head infection phase is less important on the prairies than in the eastern provinces.

Initial infections originate from soil-borne spores. Seedlings become infected following germination and further infections continue throughout the growing season. Abundant spore production occurs on diseased tissue. Spores are spread by wind, water, cultivation and through infected seeds. Spores may remain viable in soil for several years until stimulated to grow by the presence of a host plant.

Cold soils, as well as drought and high temperature stress are important predisposing factors for an infection.

The strain of C. *sativus* that affects barley is different from the strain that affects wheat. Thus, the severity of root rot may be high in wheat and low in barley in a field where a root-rot susceptible wheat cultivar was grown continuously. The reverse is also true where barley was grown continuously. Thus a crop rotation, even between cereals, will reduce root rot problems.

Damage description

Patchy emergence is usually the first indication that damage has occurred. Brown spots appear on the roots, subcrown internodes (that part of the plant between the seed and crown) and leaves. Infected plants may be stunted. Seedlings may die before or soon after emergence even though they only show slight damage.

Common root rot and crown damage lower yields. Infected plants tend to produce fewer tillers with smaller and fewer seeds per head. Some plants may ripen prematurely. Root rot in wheat and barley in Alberta causes average annual yield losses of 6 and 10 per cent, respectively.

Cereal crops can withstand a certain amount of seedling blight. The extra space, water, nutrients and light that would have been used by missing seedlings are taken up by neighboring plants to produce more tillers with larger heads and kernels compensating for the loss. No definite threshold levels are available.

Diagnosis

The presence and severity of *C. sativus* on common wheat can be determined by pulling up plants and examining the crowns and subcrown internodes for disease.

Collect 100 plants, composed of randomly selected groups of 5-10 plants each, and rate the disease level on the crown and subcrown internode by the following scale:

Clean (0) = no brown spots (lesions)

Slight (a) = up to 25% of area with lesions

Moderate (b) = 25-50% of area with lesions

Severe (c) = more than 50% of area with lesions

Use the formula (a+2b+4c) / 10, where a, b, c are the number of plants that are rated as slight, moderate and severe, respectively, in the 100 plants sampled. This gives an estimated yield loss as a percentage. Thus, if there were 25 plants in each of the slight, moderate and severe categories, the yield loss estimate equals 17.5%, calculated as follows:

$$25+2(25)+4(25)\ /\ 10 = 25+50+100\ /10 =$$
$$175\ /\ 10 = 17.5\%$$

A modification of this method only uses two disease categories: clean to slight and moderate to severe. The percentage of plants in the second group is the rating. This multiplied by 0.4 equals the estimated percentages yield reduction. If 40 per cent of the plants are in the moderate to severe category, then the yield loss would equal 0.4 x 40% or 16%.

These methods may also be used to obtain a disease rating for barley, durum wheat and rye, but no specific data are available for guidance in the conversion factor used to estimate yield loss.

Management strategy

— Avoid continuous wheat or continuous barley.
— Use rotations with other cereal or non-cereal crops.
— Avoid deep seeding, which reduces plant vigor; a healthy vigorous plant is more likely to avoid or cope with root or crown infections.
— Apply adequate fertiliser especially phosphorus to promote root growth and reduce disease severity.
— Use quality seed that was grown the previous season.
— Turn under of stubble to help reduce infection levels. Use resistant varieties.

Barley Scald

Scald is the major foliar disease of barley in the wetter regions of the prairies, particularly the parkland zone of Alberta. This disease can also affect rye and some grasses.

The scald fungus overwinters on barley residue, grain and grasses, particularly species of *Bromus*. In spring, spores are produced mainly from barley residue left on the soil surface from the previous year. Spores produced on infected leaves are transported to other barley plants by rain drops and wind. Scald can attack a barley plant at any time but levels of infection are usually most severe just before and during heading. The significance of scald-infected seed as a source of disease spread in barley is not well understood.

The disease is favored by cool (12-20 degrees Celsius), humid weather and dense crop canopies where leaves remain wet for prolonged periods.

Damage description

Scald is primarily a foliar disease but is also found on leaf sheaths and glumes. Large water-soaked, grey-green spots appear on the leaves, rapidly dry out and become bleached with brown margins. Spots often join and kill the entire leaf.

Yield losses in Alberta have been calculated at 2.4 per cent, although losses in particular fields may exceed 25 per cent. Losses are due to a decrease in photosynthetic area on the flag and second leaves resulting in reduced seed weight. Severe scald infections can cause damage that resembles black point but is not confined to the embryo end.

In general there is a direct relationship between the amount of disease present on upper leaves of cereals and the resulting loss in grain yield. The top two leaves of the cereal plant supply most of the energy required to produce well-filled grain. This principal can be used to predict potential yield losses at the milky ripe growth stage.

If scald appears on the upper leaves and sheaths in mid-July, anticipate considerable yield loss. If the scald does not appear until early to mid-August, disease levels may be heavy but actual losses from this disease will be considerably reduced since the grains are well filled by this time.

Diagnosis

Examine the crops at the milky ripe stage (see Feekes scale). Assess no less than 25 main tillers selected at random along two diagonals from one corner to the opposite corner of the field. Assess the percentage of first and second leaves affected. Take the average of the first leaf and the second leaf and apply it to the following formula. This will give you a reasonable estimate of expected crop loss from scald.

% yield loss = 2/3 x % area of flag leaf infected + 1/2 x % area of second leaf infected / 2

Management Strategy

— Use resistant cultivars. Turn under or deep till surface barley residue. This reduces disease levels when barley follows barley.
— Seed early with an early maturing variety to miss the major build-up of disease that hits later-sown crops and later maturing varieties.
— Use a crop rotation of at least one year with non-host crops such as other cereals or canola.
— Use resistant cultivars.
— Apply a foliar fungicide. Tilt 250E/Bumper 418EC (propiconazole) and Headline EC (pyraclostrobin) are registered on barley in Canada for control of scald and net blotch.
— Baytan 30 (triadimenol) applied as a seed treatment will provide suppression of the seed-borne phase of scald.

Net blotch

Net blotch is a common, destructive foliar disease of barley in western Canada, second only to scald as a cause of yield loss. Net blotch is more common than scald in the warmer, drier regions of the prairies.

The fungus overwinters on the seed or crop residue. Spores are produced and spread by wind and rain. Spores produced on infected plants are mainly responsible for destructive, secondary spread of the disease within the crop.

Seedling infection is greatest during periods of cool humid weather (10-15 degrees Celsius). Spore production and infection of the growing crop is favoured by high relative humidity and temperatures around 20 degrees Celsius.

Damage Description

Light brown spots with distinctive dark brown net-like patterns appear on the leaves, sheaths and glumes. Spots enlarge and join into dark brown stripes. There is also a spot-producing form of this fungus that produces dark brown spots surrounded by a yellow zone. Symptoms vary with the variety and weather.

Yield loss is proportional to the amount of leaf area destroyed on the upper two leaves. The average yield loss in Alberta is about 1.5 per cent. Losses of 50 per cent and more have occurred in highly susceptible varieties such as Elrose. Net blotch lowers grain yield and brewing quality by reducing the carbohydrate content of kernels.

Diagnosis

Use the barley scald assessment procedure to obtain an indication of crop loss.

Management Strategy

— Use resistant cultivars. Use resistant cultivars.

— Bury crop residue and destroy volunteers.

— Use balanced applications of nitrogen and phosphorus. Heavy nitrogen applications generally produce conditions favourable to outbreaks of this disease.

— Follow a crop rotation for at least two years with non-susceptible hosts. Barley should not follow barley, particularly if disease levels were high the previous year.
— If barley must be grown in two successive years, use a susceptible cultivar the first year and a resistant type the second.
— Use disease-free seed if possible. Seed may be sent to the Agriculture Canada Seed Laboratory in Ottawa to determine the percentage of infection.
— Control the seed-borne phase of net blotch with a chemical treatment.
— Apply a foliar spray. Tilt 250E/Bumper 418EC (propiconazole) and Headline EC (pyraclostrobin) are registered on barley in Canada for effective control of net blotch and scald.

Barley Stripe

Barley stripe can cause significant yield losses in barley, other cereals are not affected.

This fungus is exclusively seed-borne and overwinters on or in the seed. The fungus grows internally within the infected plant. Spores are produced on the infected leaves and are spread by wind to nearby healthy heads. The seeds become infected only in the field and are most susceptible during early development of the seed head.

Infection of seedlings is higher when soil temperatures remain below 15 degrees Celsius. Sporulation on the foliage of infected plants is encouraged by high humidity.

Damage Description

Vivid longitudinal yellow stripes bounded by the veins appear on the leaves and extend from the base of the leaf to the tip. The stripes eventually become brownish and infected leaves characteristically tear and fray. Infected plants are stunted and heads become twisted, blighted or fail to emerge.

The yield loss is directly proportional to the percentage of infected plants. A 1 per cent infection causes a 0.7 per cent direct yield loss. Some crop compensation occurs from neighbouring healthy plants.

Diagnosis

Infected plants are most obvious just before heading out, and the disease can be readily identified at this time. After heading out, infected plants may be "lost" under the canopy. Seed samples may be sent out to Agriculture Canada Seed Laboratory in Ottawa to determine the percentage infection.

Management Strategy

— Use disease resistant varieties.

— Avoid infested seed.

— Use a fungicidal seed treatment

All cereals are attacked by smut, but each crop is host to specific species of smut fungus. Each smut species has a number of races that differ only in their ability to infect certain cultivars of that crop host.

Covered and false loose smut of barley, bunt of wheat and loose and covered smut of oats will be dealt with as a group because their disease cycle is similar. These fungi overwinter as spores on the seed surface. When infested seeds are sown, smut spores germinate, penetrate the seedling and grow within the cereal host until the heads develop. Smut fungi replace all or most of the grain head and form masses of black smut spores instead of seeds and chaff. These spores are released at grain harvest and contaminate the surface of other healthy kernels.

True loose smut of barley and wheat differs from the above cycle in that the fungi overwinter within the embryo of the seed. The fungus grows within the plant and produces loose black smutted heads. Wind-borne, dust-like spores infect healthy cereal flowers, where they penetrate the developing seed and infect the germ (embryo).

True loose smut infection is favored by cool wet weather in July, which slows and prolongs the flowering period and allows more time for infection. All other smuts are favored by seeding into cold soils during the late fall or early spring.

Damage Description

Most smut fungi attack and replace the internal tissues of the grain with dark brown or black smut spores.

— *Covered smut (barley, oats)*The plant may be slightly stunted with hard, compact, upright, smutted heads. The smut balls are covered in a membrane that remains intact along with the awns and chaff.

— *False loose and true loose smut (barley):* These fungi have a very thin covering membrane around the spore masses. The brown spores are blown or washed away and leave bare spikes in the ripening crop. Losses from loose smut in barley average less than 1 per cent, but losses of up to 40 per cent have been recorded. A laboratory examination is needed to tell them apart.

— *Bunt (wheat and triticale):* Infected heads may be wholly or partially smutted. The heads may remain green longer and glumes in the infected head may be spread apart. Awns may be reduced or fall out as the heads ripen. Bunt balls are rounder than wheat kernels. When these balls rupture, the loose black powdery spores have a distinctive fishy odor.

Smuts attack the grain directly, replace the contents and cause a 1:1 direct yield loss. A 10 per cent bunt infection, if the heads are completely infected, results in a 10 per cent yield loss (aside from quality considerations). The presence of covered and false loose smut and bunt also results in a drastic reduction in grain quality because of the visible contamination with black smut spores.

Smutted grain should be stored separately from clean grain. Heavily smutted or bunted grain will not be accepted at the elevator and may even be difficult to sell as feed owing to respiratory or feed refusal problems that might result. Smuts and bunts are not toxic to livestock.

Economic Threshold

The maximum percentage of smutty barley kernels (true loose smut) permitted under the Canadian Seeds Act, July 1, 1987, are as follows:

Grade	*Maximum % of smutty barley kernels*
Canada Foundation #1	2
Canada Foundation #2	4
Registered #2	4
Certified #1	2
Certified #2	4
Common #1	4
Common #2	6

The fishy odor from bunted wheat is detectable at less than 0.01 per cent thus rendering grain unsuitable for food even at this very low level of contamination.

Diagnosis

Use the following formula to calculate losses from loose smut and bunt in barley, wheat and oats. The percentage of infected ears is directly proportional to grain yield loss.

Loss (kg/ha) = Pr x Pa / 100-Pr

Pr = percentage infected heads

Pa = actual yield (kg/ha)

This is done during heading for loose smut and true loose smut and during ripening for bunt. Assess 50 plants along a diagonal at sites selected at random. Calculate the percentage of infected heads. Thus a 10 per cent infection in a 3000 kg/ha crop = 333 kg/ha yield loss.

Management Strategy

— Use resistant cultivars.

— Use smut free seed.

— In southern Alberta bunt of wheat may be soil-borne. Avoid continuous cropping of winter wheat and do not seed winter wheat within 2 km of a known bunt infected field of either spring or winter wheat because the spores are wind blown.

— Use a hot water seed treatment for loose smut. Dip seeds in 20 degrees Celsius water for 5 hours, drain and then dip in 49 degrees Celsius water for 1 minute, then 52 degrees Celsius water for exactly 11 minutes, then immediately in cold water until seeds cool off. Some seed may be killed by hot water treatment.

— Use systemic seed treatments that contain carbathiin for loose smut control in wheat and barley. All other semi-loose, covered smut and bunts in cereals are controlled by any of the registered fungicide or fungicide combinations.

False Stripe

Stripe mosaic is a seed-borne virus. In Canada, the virus has occasionally been detected in pedigree seed, but losses caused by BSMV have been rare in recent years. Barley is the main host. On occasions it has been found in wheat and several grass species.

The virus is seed-borne and is transferred from plant to plant when crop leaves rub against one another. Experimentally the virus can be transmitted by pollen, but since barley is self-pollinated this method of spread is generally of no consequence. Infected seeds produce infected plants. Seed from virus-infected plants is generally infected to a 60 per cent level.

Disease builds up when infected seed is planted year after year.

Damage Description

Symptoms may vary with the virulence of the BSMV strain and time of infection. Infections appear as chlorotic mottling with spots or stripes of a yellowish color. Infected plants may be stunted and may mature later than healthy plants.

Yield losses are proportional to the level of infection in the seed lot. Losses are caused by reduced grain production, fewer heads per plant, semi-sterility and incomplete head emergence from the sheath. Heavily infected crops have had yield reductions of up to 25 per cent. The percentage of infected seedlings indicates the level of grain infection.

Diagnosis

For a positive identification of the disease send infected (mosaic symptom) seedlings to your nearest plant disease laboratory along with some seed from the seed lot that was sown.

Management Strategy

Avoid introducing the disease. Montana requires seed from Canada to be certified free of this disease. Seed destined for Montana must be tested at the Montana Seed Laboratory, Montana State University, Bozeman.

— Use virus-free seed.
— Control volunteer barley. Do not plant barley after barley if the previous barley crop was infected.

Barley Yellow Dwarf

The barley yellow dwarf virus (BYDV), also called red leaf in oats, can infect barley, oats, rye and wheat as well as numerous species of grasses. It occurs in most parts of the world and is considered the most common viral disease of cereal crops. BYDV is transmitted by several species of aphids and occurs in many strains or types.

Usually the aphid vectors cannot overwinter in Alberta and they normally drift in every spring from the United States. These aphids may already be infective they but can also pick up the virus from infected perennial grasses in Alberta. As the aphids (winged or wingless) feed on the cereal crop, they transmit the virus through their mouth parts. The aphids can remain infectious for life, which is around 40 days.

Damage Description

Symptoms vary with the host species and the stage of crop development. Infections at the seedling stage may result in death or dwarfing as well as sterile heads.

Leaves turn yellow from the tip down, along the leaf margins or in blotchy patches. Infected barley leaves, particularly flag leaves, turn bright yellow; in oats, the leaves may turn from red to purple. Discolored areas enlarge and progress to the base of the plant. Heads may be wholly or partially sterile. There may also be an increase or decrease of tillers produced by infected plants. Cereal plants infected early in the season may be shaded out by healthy or late infected surrounding plants. Winter wheat seedlings may be 100 per cent infected with BYDV before freeze-up in the fall.

In dry seasons in the Fort McCleod/Claresholm area, barley that is seeded in June may turn yellow in early July. In alkaline soils this symptom indicates a manganese deficiency in the cereal crop.

BYDV affects yields by stunting, reduced tillering, sterility, and failure to fill kernels. However, because the aphids must move in from the United States, they generally arrive too late in the season to cause significant yield loss in Alberta. In some regions in the United States, yield losses reach 30-50 per cent in oats and 5-30 per cent for barley and wheat. On the Prairies, the only crops that are damaged significantly are late sown spring crops and early sown, fall-seeded crops.

Bright yellow flag leaves in barley by late June or early July will indicate little or no yield contribution from those plants. Severe stunting in winter wheat by mid-May could indicate crop failure from one or more strains of BYDV.

Diagnosis

Visual symptoms will indicate the presence of the disease. Confirmation of this virus must be carried out at a research facility.

Management Strategy

— Use resistant varieties.
— Seed early in the spring. This will allow for maximum growth of the cereal crop before possible infection by migrating aphids.

Insect Pests of Barley

Damage from insect pests can usually be minimised by cultural practices and proper use of insecticides. Cultural control should be used wherever possible to reduce the need for insecticides. When insecticides are necessary to prevent loss, they should be applied with caution and only as recommended.

Wireworms

Wireworms are slender, hard-bodied, shiny, yellow worms, the largest about 2.5 cm (one inch long). Eggs are laid by adult beetles in the spring and larvae may remain in the soil for 5 to 10 years before pupating and emerging as adult beetles in early summer. Populations build up slowly but fields remain infested for years if larvae are not controlled. Wireworms feed on germinating seeds and underground parts, shredding the stems but seldom cutting them off. Severe damage results in thin patchy stands. The central leaves die but outer plant leaves often remain green for some time.

Barley is more tolerant to wireworm damage than wheat or rye, but will be damaged when infestations are heavy, such as on new breaking of grassland. Current seed treatments do not control wireworm. Shallow seeding into moisture and firm packing will effectively prevent damage when populations are low.

Insect Life Cycle

- *Host plants:* The prairie grain wireworm is the most destructive wireworm pest of grain in Western Canada. It prefers annual and perennial grasses. It also attacks potatoes, sugar beets, corn, lettuce, sunflower, canola and onions.
- *Overwintering:* Wireworms (the larval stage) live for several years in the soil and are quite resistant to adverse conditions. Those larvae that survive their first winter can go for at least two years without any food other than humus. The wireworm stage lasts anywhere from four to 11 years. They hibernate in the soil from 5 to 25 cm below ground level. Older larvae commonly

feed to a depth of 15 cm in the topsoil. When fully grown, usually in July, the larvae pupate about 5 to 10 cm below the soil surface. Pupation lasts for less than a month; however, adults do not emerge until the following spring.

— *Spring appearance:* Wireworm adults, called click beetles, emerge in April and early May from the soil in which they overwintered. They become active when the air temperature is above 10 degrees Celsius, mate and then seek egg laying sites. From late May through June individual females deposit 200 to 400 eggs in loose soil, or under lumps of soil. Depending on the moisture, temperature and firmness of the soil, eggs are laid anywhere from just below the soil surface to 15 cm deep. After three to seven weeks, the wireworms hatch and begin to feed on live roots or seeds of cereals or grasses. If no food is found within one to four weeks of hatching, the larvae die.

— *Number of generations:* The generation time (4 to 11 years) varies with the quality and availability of food. Wireworms in all growth stages are likely to infest a field.

— *Natural enemies Parasites :* Parasitic nematodes attack larvae. Pathogens - Larval mortality caused by bacterial diseases is especially high in soils with a high moisture content.

Damage Assessment

Economic importance

In Alberta, damage to wheat crops ranges from one to 50 per cent annually. Damage to other non-cereal crops varies from farm to farm. Over 80 species of wireworms occur in Alberta, but only the prairie grain wireworm is of major economic importance.

Damage description

The larvae feed on germinating seeds or young seedlings and shred the stems but seldom cut them off. The central leaves die but outer leaves often remain green for some time. Damaged plants soon wilt and die, resulting in thin stands. Poor seed and dry conditions can also cause thin stands; consequently, many wireworm infestations are passed off as due to poor seed or poor germination. Wireworms do the most damage in early spring when they are near the soil surface. During summer months, larvae move deeper into the soil where it is cool and moist. Wireworms do not ingest solid plant material, but chew tissues, regurgitate fluids containing enzymes, and then imbibe the juices and plant products made soluble by the enzymes.

Potato seed pieces are seldom damaged to a point where poor stands result. However, new tubers can be damaged severely. Tunnels made by the wireworm allow disease organisms to enter, and the damaged tubers are less marketable.

Damage is generally higher in silty, medium textured, well-drained soils and in soils cultivated for at least 12 years. Damage is less likely in heavy or very light soils. Crops grown in newly broken sod can suffer great losses for one to two years, then the damage decreases rapidly only to gradually increase in succeeding years if no wireworm control measures are applied.

Sampling and monitoring

Whole potatoes buried in marked locations in a field in the spring or from early to mid August will indicate whether wireworms are present. Bury the potatoes 10 to 15 cm deep then dig them up after a couple of weeks, and examine them for wireworm tunnels. Monitor your fields each year. To sample for larvae, sieve the soil through a screen. Mark out areas 50 cm x 50 cm and sieve the soil to a depth of 15 cm. Repeat in different areas of the field to determine an average number of larvae per square metre.

Economic threshold

None have been established. Treat seed for two consecutive years after breaking sod to reduce the problem to a non-economical level.

Management Strategy

Effects of weather

Larval activity is governed by temperature and moisture conditions. Cool wet weather forces wireworms closer to the surface; dry hot weather forces them deeper into the soil. Cool weather restricts adult activity and lengthens the egg laying period. Eggs laid near the soil surface or in compact soil are subject to high mortality when moisture levels and temperature fluctuate rapidly. Mortality is from 92 to 98 per cent in eggs and young wireworms. Most wireworm mortality occurs during the first two weeks of larval life.

Cultural practices

Crop rotation - Crop rotation and other cultural practices usually prevent wireworms from becoming a major problem in sugar beet fields. Because sugar beets are normally grown in a four-year rotation in Alberta, crops less susceptible to wireworm attack can be grown on infested fields so that populations will not build up. Root and row crops such as potatoes, corn, onions or beans should not be grown in a rotation where wireworms have been a problem. Wireworms can also be present when sugar beets are grown on land previously uncultivated or planted to grass or pasture. Deep plowing in the fall and frequent cultivation in early summer are suggested when wireworms are known to be present in these fallow fields.

Shallow cultivation - The long life and underground habit of wireworms make them hard to control. The most vulnerable period for the wireworm is from the egg to early larval stage. Only two to eight per cent of eggs and young wireworms survive. Newly hatched wireworms must feed within four weeks. Thus, cultivate fallow land in early spring to starve hatchlings. Shallow cultivation in early spring can expose eggs and injure larvae. Thorough cultivation of summer fallow during the latter half of July can destroy pupae as well as larvae. Use a rod weeder and disturb only the upper soil layer.

Summer fallow - Summer fallow for wireworm control is not recommended, because it has almost no effect on mature larvae, which can survive for two years on soil humus alone. In fact, wireworm damage is more severe after fallow. If early spring cultivation is used to starve and injure young larvae and eggs, plant a resistant crop.

If summer fallow must be part of the rotation, starve newly hatched wireworms by destroying all green growth during June and July. Work summer fallow as shallow as possible for weed control. Seed shallow, pack the seedbed to induce quick germination, and avoid very early or very late seeding.

Shallow cultivation or seeding combined with soil packing -Pack behind the seed drill to reduce damage to grain in wireworm infested land. Wireworms are very poor travellers. Some remain in the larval stage for nine or 10 years during which time they travel only a few yards. Firming the soil further impedes wireworms. Use a press drill for best results. If a press drill is not available, use a packer hitched behind the seeder in such a way that all wheels of the packer "follow" the drills of the seeder. Thus, the seed row is packed firmly, making wireworm movement so difficult that most of the worms will seek their food in the looser soil between the seed rows. If the packers do not follow the seed rows they will tend to leave them loose while firming the intervening strips. This may encourage the worms to follow the seed rows and cause heavy damage.

Farmers should restrict tillage operations to the upper 5 to 8 cm of the soil to maintain a compact soil layer beneath the tilled layer.

Adult click beetles are also affected by a compact sub-layer because they are forced to lay their eggs close to the surface, where the eggs can easily dry out or be discovered by predators.

Seeding practices - Avoid very early or very late seeding. Use methods that speed germination and early growth of the crop to help reduce the impact of wireworm damage. Cultivate before spring seeding to give the crop a competitive advantage with weeds and to prepare a good seedbed. On land that was in summer fallow the year before, a rod weeder will produce a very compact seedbed. On stubble, use a one-way disc; then seed at the proper time. If the soil is moist, seed 10 to 14 days from when drilling was first possible so that the soil is warm and seeds will grow quickly. If the soil is dry, delay seeding until it rains. As much as 90 to 95 per cent of a crop has been destroyed when seeded into a dry soil compared with 5 to 10 per cent loss when seeded into moist soil. Moisture also helps young seedlings recover from wireworm damage.

Shallow seeding - Seed preferably at a depth of 2 to 5 cm. This speeds the early growth of plants (as long as moisture is present). Use a drill press or standard drill with press attachment. If not available, use a standard double disc drill. Avoid the use of single-disc drills, and hoe-drills; but if you must use them, use press attachments or follow with a packer.

Increase the seeding rate in fields infested with wireworms, especially for wheat. Use as much as an extra bushel per acre, or for a patchy infestation, drill twice. Use healthy seed. A light topping of rotted manure or an application of phosphate fertilisers will help reduce wireworm damage. The phosphorous in the manure probably encourages root development and early maturity. Manure can be applied late in the fallow year or early the next spring, and should be incorporated into the soil.

Do not plant susceptible crops on the same land in two successive years. A crop rotation with resistant varieties and legumes is useful.

Buckwheat and flax are not usually damaged by wireworms. If recommended seeding practices are followed, oats and barley can be planted except when infestations are severe. Fall rye and winter wheat are more resistant owing to early vigorous growth in spring. However, a wet autumn preceded by a dry summer and fall has resulted in damage to rye. Plant legumes with a light nurse crop or in a mixture with a grass. Seedlings of sweetclover and alfalfa can be seriously damaged by wireworms. The seedlings usually escape damage in a mixture. Mature plants are not damaged by wireworms.

Take special precautions after breaking sod. If a crop on new land is destroyed, reseed immediately with a resistant crop. This is preferable to leaving the land fallow, since a recurrence of the problem would then be likely.

— *Biological control Parasites*—A few nematode parasites control wireworms that pupate below ground. At present these nematodes are not available for commercial use.

— *Pathogens* - Wireworms are susceptible to bacterial and fungal diseases, and mortality is higher in moist soils.

— *Predators* - The adults are preyed upon when they lay eggs, the larvae are eaten in spring when they are near the surface or when exposed by cultivation. Birds pull them from the soil, and moles and shrews dig for them.

Cutworms

Cutworms are fleshy, soft-bodied worms that curl up and remain motionless when disturbed. Fully grown, they are about 3 to 5 cm (1.15 - 2 inches) long. The most common species are the pale western cutworm, most prevalent in grassland areas, and the redbacked cutworm, prevailing mainly in parkland areas. Both feed at night, cutting off plants just at or below the ground surface. Damage is most severe when plants are in the seedling or early tillering stages.

Damage by both species can be controlled by discouraging egg laying. Leaving summerfallow undisturbed during August to mid-September reduces egg laying by pale western cutworm moths which lay at this time into loose and dusty soil. The redbacked cutworm will lay in weedy fallow and weedy patches in crops, therefore, weedy summefallow in parkland areas should be worked during this period. Young larvae of both species can be starved in the spring by destroying green growth 10-14 days before seeding. Because late seeding will result in lower yields, it is usually preferable to seed during the optimum time and use a chemical control if necessary for cutworm control.

Grasshoppers

Grasshoppers can damage field crops, although only three of the many species of grasshoppers in Alberta are a threat. All over winter as eggs that hatch in the spring. Depending on species, eggs may be laid on roadsides or headlands, or scattered throughout stubble fields. Damage is most severe during dry periods in the early stages of crop growth. Relatively cool and moist growing conditions reduce grasshopper feeding and encourage crop growth, allowing the barley to keep ahead of damage. Seeding early and producing a vigorous crop assists in reducing damage.

Forecasts of the expected extent and severity of outbreaks are prepared jointly by Alberta Agriculture, Food and Rural Development; Agriculture and Agri-Food Canada; and local agricultural service boards. The forecast is in most farm papers. When grasshoppers first hatch in breeding areas such as roadsides and headlands, they can be controlled by early applications of insecticides to these relatively small areas. Fall tillage helps to destroy eggs by exposing them to predators and the elements.

Insect Life Cycle

— *Host plants:* The clear-winged grasshopper is mainly a grass feeder. Economic damage is primarily to cereals, especially wheat and barley. Clear-winged and migratory grasshoppers have together destroyed areas of range grass and hay almost entirely.

The nutritional qualities of the chief food plant can affect longevity and egg production of grasshoppers. Kentucky blue grass is one of the best foods for high survival and egg production. Western wheatgrass is one of the worst foods for grasshoppers and could be a factor that limits distribution of this grasshopper on the Prairies. It may be the only green grass available to attract females at egg laying time.

— *Spring appearance:* If the previous fall was warm, eggs of the clear-winged grasshopper will be the last of the pest grasshopper species to hatch. They hatch in late May to early June. The embryos of this species undergo a maximum of 50 per cent of their total development before winter. The other pest species can complete more of their development in the fall and consequently need less time in the spring to complete pre-hatch development.

— *Overwintering:* Clear-winged grasshopper eggs are laid in the fall and hatch the following spring. Each female lays an average of eight egg pods (about 175 eggs) usually in unbroken sod. The short vegetation of dry, mowed roadsides and sparse, over-grazed pastures is especially favored for egg deposition. Males are conspicuous at egg laying sites; their undersides become bright yellow during the mating and egg laying period. They stake out a territory and wait for females to come to oviposit.

— *Number of generations:* As with all of Alberta's grasshopper pests, clear-winged grasshoppers have one generation per year.

— *Natural enemies:* Next to weather, natural enemies are the grasshopper's most important population control factor. In some localized areas natural enemies may cause even more mortality than the weather.

Some enemies attack when grasshoppers are still in the soil awaiting spring. Others attack nymph and adult stages.

Predators - Among the most important egg predators are bee flies, blister beetles, ground beetles and crickets. Common field crickets eat the eggs and may destroy up to 50 per cent of the eggs in some areas. Bee flies and blister beetles deposit their eggs in the soil near grasshopper eggs. When the larvae of these egg predators hatch, they locate the egg pods and feed upon the eggs. If bee flies and blister beetles are abundant, they may destroy up to 80 per cent of eggs in a localised area.

Spiders, some wasps and many birds feed on grasshoppers and consume large numbers of nymphs and adults. Their effect on the total grasshopper population is not known.

Parasites - A few other insects, such as the tiny wasps of the genus Scelio, parasitise eggs just after they have been laid. The young parasitic larvae complete their development within the eggs in time to emerge as adults and parasitise the eggs of the next generation of grasshoppers. They may destroy from 5 to 50 per cent of the grasshopper eggs.

Parasites of nymphs and adults include flesh flies, tachinid flies and tangled-veined flies. Fly larvae burrow into the grasshopper when they contact it on the ground. Other fly larvae are deposited on or into the grasshopper's body by the female fly. The maggot then feeds inside the grasshopper and eventually kills its host as the maggot leaves the body. This group of insects may parasitise up to 60 per cent of the nymphs and adults.

Threadworms attack grasshoppers if the young worms encounter a grasshopper or if grasshoppers eat threadworm eggs. Threadworms overwinter in soil and lay their eggs on the soil or on vegetation.

Pathogens - The fungus, *Entomophthora grylli,* can effectively control grasshoppers under warm, humid conditions. This fungus may occasionally reach epidemic proportions. The disease leaves the corpses of its victims clinging to the stems of plants.

The microsporidian parasite, *Nosema locustae,* is an effective enemy of grasshoppers. A grasshopper becomes infected if it eats contaminated vegetation or a diseased grasshopper. A grasshopper population infected with this organism may be reduced by as much as 60 per cent in one year. It also reduces the number of eggs laid and restricts the movement of individuals and thus affects grasshopper populations. Perhaps this organism's greatest potential as a biological control agent, however, is for reducing food consumption.

Since most of the natural enemies of grasshoppers are already widespread, it is unlikely that they could be used to prevent grasshopper outbreaks over extensive areas.

Nevertheless, natural enemies do control localised grasshopper infestations and hasten the decline of grasshopper outbreaks.

Poultry—During the depression, some farmers successfully used turkeys to control grasshoppers. In years when grasshoppers were plentiful, the turkeys were simply released into fields. This resulted in a secondary benefit: the turkeys required little supplemental food since grasshoppers provided a plentiful high quality protein diet.

Other birds—Gulls, hawks, crows, meadowlarks, crowned larks, lark buntings, desert horned larks, shrikes, curlews, killdeer, partridges and cranes are all predators of grasshoppers. Many birds scratch up the egg cases, and have been credited with clearing from 5 to 150 acres of grasshopper pods. Birds, especially gulls and meadowlarks, are credited with stopping some infestations in the early part of the century. This is more likely to happen if a habitat provides food and refuge for the birds. Farm habitat can be made more attractive to birds and so encourage predation of insect pests.

Vertebrates other than birds—Most animals are opportunists; they will eat what is nourishing and available to them. Mice, rats, shrews, gophers and badgers all eat grasshoppers and their egg pods. Coyotes, skunks, lizards, snakes, toads, bobcats and kit foxes eat nymphs and adults.

Damage Assessment

Economic importance

Clear-winged grasshopper has been on average, the most economically important species of grasshopper in Canada, although over the years its importance relative to other species has changed gradually. The first record from the Prairie Provinces was from Saskatchewan in 1800. Clear-winged grasshoppers did not become abundant, however, until about 1900 when road-building, drainage and cultivation in southern Manitoba created favorable breeding sites and greatly increased the abundance of suitable food plants. In recent years there has been a marked decrease in the intensities of outbreaks in Manitoba. This may be related to changes in agricultural practices that have reduced populations of certain grasses.

Damage description

Damage to cereal crops is generally concentrated near field margins and is caused when hatchling grasshoppers move out of egg beds into field edges; damage to grasslands tends to be more evenly distributed. Damage to cereals includes leaf notching and stripping but is most costly when stems are severed just below the heads of maturing or mature crops. When grasshopper numbers are extremely high and natural plant hosts in short supply, grasshoppers will consume or attempt to consume any plants or plant products that they come upon during their migrations in search of food.

Sampling and monitoring methods

Walk through the infested area and estimate the number of grasshoppers per square metre as they jump in front of you. A sampling 'T' will likely improve your estimate. The 'T' consists of a metre-long measuring stick, carried by a handle so that a square metre can be visualised at crop height. Walk and carry the 'T' just above the crop.

Late summer and fall surveys of grasshopper adults have been carried out by agriculture fieldmen in Alberta since 1932. Grasshopper forecast maps are produced yearly from data collected in about 1,700 townships.

Spring surveys of grasshoppers and grasshopper eggs are also conducted in years when high grasshopper numbers are expected. In this way, improved estimates are obtained for time of hatch, population density and the effects of predators and parasites.

Management Strategy

— *Effects of weather:* Population size is primarily determined by weather. Outbreaks are usually preceded by two to three years with hot, dry summers and open falls. Dry weather increases the probability of egg survival, hastens spring hatch, and promotes nymphal development and adult feeding. Open falls allow grasshoppers more time to feed and lay eggs. Cool, wet weather increases egg mortality by promoting fungal diseases, retards nymphal development, reduces the numbers of eggs laid by delaying sexual maturity and reduces the activity of grasshoppers at all stages.

— *Cultural practices:* Tillage - Cultivation is probably the most effective cultural practice available to farmers for the reduction of grasshopper populations. Tillage controls grasshoppers primarily by eliminating the green plants on which grasshoppers feed.

However, tillage is of little value for the sole purpose of physically destroying grasshopper eggs or exposing them so that they dry out or are eaten by birds and other insects. Excessive tillage may also increase the risk of soil erosion.

Tillage to eliminate weeds from summer fallow fields during late summer and early fall will discourage female grasshoppers from depositing their eggs in these fields. Grasshoppers seldom lay eggs in clean summer fallow even when it has a heavy cover of trash. Similarly, thorough cultivation of fields immediately after harvest will help discourage grasshoppers from laying all their eggs within the field.

Complete spring tillage before grasshoppers hatch to eliminate all green growth on stubble fields that are to be in summer fallow. If no food is available when grasshoppers hatch, they will starve to death because they are unable to move long distances to find food. Early tillage will also give good weed control and conserve moisture at no extra cost.

Trap Strips - If grasshoppers are present when tillage operations begin, elimination of all green plant material in a field will probably not achieve adequate control. Once grasshoppers have fed and developed to the second stage of growth (second instar), they usually are mobile enough to move to adjacent crops when their food supply becomes exhausted. Use trap strips in these fields to collect grasshoppers into a relatively small area where quick and economical control will be possible with a minimum of insecticide.

To make strips, cultivate a black guard strip 10 m wide around the outside of a field. Leave an uncultivated green strip at least 10 m wide before resuming cultivation. Repeat the process as often as necessary to produce additional trap strips. All green vegetation must be eliminated between the trap strips if they are to be effective. The black guard strip will ensure that grasshoppers promptly move into the trap strips to feed.

The effectiveness of trap strips can be improved considerably by seeding them to wheat, barley or oats several weeks before tillage begins. Trap strips should have adequate vegetation to feed even the largest of grasshopper populations for three to five days.

Migration of young grasshoppers from the cultivated guard strips to the trap strips may take several days. Once the migration is complete, the trap strips and a 10 m strip of adjacent crop should be treated with insecticide. Apply the highest recommended rate of insecticide to ensure adequate control.

Before cultivating the trap strips, allow three days to assess the effectiveness of the insecticide. If adequate control is not achieved after three days, treat the trap strip again. When grasshoppers have been eliminated from the trap strip, it should be possible to complete tillage without fear of displacing large numbers of grasshoppers into adjacent crops.

Summer fallow that is not properly managed can be a major source of grasshoppers. Many cases of growers applying insecticides six or seven times to a field border are a direct consequence of improper grasshopper control on summer fallow.

Early seeding—Seed crops as early as possible. Older plants can withstand more grasshopper damage than younger plants that are not well established. Although early seeding may not totally prevent crop damage, damage will be reduced and more time will be available to apply insecticides. Also, crops that are seeded earlier will mature earlier and migrating grasshoppers are not as likely to be attracted to them.

Crop rotation—Whenever possible, avoid seeding cereals on stubble fields heavily infested with grasshoppers. Seed cereals only on stubble fields where soil moisture is adequate and where one or more applications of an insecticide over the entire field is economical.

Canola seeded with an in-furrow application of a granular insecticide may be a suitable alternative to cereals where grasshopper infestations are light.

Roadside vegetation management—Certain of our common roadside weeds, such as stinkweed, alfalfa and dandelion, are nutritious food plants for nymphs and adults. Such plants promote high survival and egg laying. Western wheatgrass on the other hand, is one of the poorest food plants.

Aphids

Aphids seldom inflict economic damage on barley crops in

Alberta. Infestations do not normally build up to damaging levels until the crop is nearly mature. Damage occurs most frequently when barley is sown late in the spring or as a cover crop in the fall. If very heavy infestations build up (50 per plant) before heading, chemical control may be justified. Of the many species present, only 3 or 4 are of economic importance.

BARLEY THRIPS

Adult barley thrips are very small, less than 2 mm long. To the naked eye they appear slender, dark coloured and shiny. Females are winged and males are wingless. The wings lay flat along the back and are visible as pale or silver strips. The larvae are wingless, pale yellow, white or green with a slightly darker head capsule and red eyespots. They have two pupal stages. The first ("prepupal") stage lasts only a few hours to one day; the pupal stage lasts from two to six days.

Host plants

Barley thrips overwinter in sod along roadsides and in shelterbelts. They prefer Kentucky bluegrass and brome sod in shelterbelts. During the growing season barley thrips can be found in barley grown throughout western Canada. In the fall, females migrate from the crop into overwintering sites. Males and late maturing larvae remain on the crop but are killed by starvation and early frosts.

Overwintering

Only the adult females overwinter in debris in fields and along headlands, especially near shelterbelts, in sod or under bark. They have also been collected from forest litter, leaf mould, insect galls and birds' nests.

Spring appearance

In early spring, thrips leave their overwintering sites by

launching themselves from the tip of a leaf or grass awn and flying short distances or being carried long distances on wind currents. Research shows that there is often a flight in early spring, probably to grass hosts, and another flight later in spring to cereal hosts. Once in a crop, flight is limited to very short distances between host plants. Females mate in the fall but require approximately six weeks in the spring to mature eggs. Eggs are then laid, either scattered or in short rows, in an incision usually made alongside the leaf veins. Upon hatching, the larvae feed for approximately seven days, shedding their skin twice. Adult male barley thrips mate with the pupal stage of the females. Therefore, females are already mated when they emerge from the pupae. After two or three days adults emerge and feed until the crop begins to mature.

Damage assessment

Barley thrips infest barley at a very early stage and remain feeding on the inside of the leaf sheath. At the beginning of head formation, intensive feeding produces heads that appear white or silvery and may produce thin, shrivelled grains. Often there is no seed development at the top and bottom of the head and intermediate grains are shrivelled. Yield losses may also occur if feeding in the sheath is intensive enough to kill the flag leaf. Without the flag leaf, kernels do not fill properly and seed weight is reduced.

REFERENCE

Brunt, A.A., Crabtree, K., Dallwitz, M.J., Gibbs, A.J., Watson, L. and Zurcher, E.J. (eds.), *Barley mild mosaic bymovirus*, 1996.

5

Insects Pests of Oats

The common oat plant (*Avena sativa*) is a species of cereal grain grown for its seed, which is known by the same name (usually in the plural, unlike other grains). While oats are suitable for human consumption as oatmeal and rolled oats, one of the most common uses is as livestock feed. Oats make up a large part of the diet of horses and are regularly fed to cattle as well. Oats are also used in some brands of dog and chicken feed.

Oats are grown throughout the temperate zones. They have a lower summer heat requirement and greater tolerance of rain than other cereals like wheat, rye or barley, so are particularly important in areas with cool, wet summers such as Northwest Europe, even being grown successfully in Iceland. Oats are an annual plant, and can be planted either in autumn (for late summer harvest) or in the spring (for early autumn harvest).

Historical attitudes towards oats vary. Oat bread was first manufactured in England, where the first oat bread factory was established in 1899. In Scotland they were, and still are, held in high esteem, as a mainstay of the national diet. A traditional saying in England is that "oats are only fit to be fed to horses and Scotsmen", to which the Scottish riposte is "and England has the finest horses, and Scotland the finest men".

PESTS AND DISEASES

Oats are relatively free from diseases and pests, with the exception being leaf diseases, such as leaf rust and stem rust. A few Lepidoptera caterpillars feed on the plants—e.g. Rustic Shoulder-knot and Setaceous Hebrew Character—but these rarely become a major pest. See also List of oats diseases.

Leaf Rust

Leaf rust is the most serious problem affecting grazing oats in Queensland. Leaf rust is a disease caused by the fungus Puccinia coronata avenae. Growing a leaf rust resistant oat variety is the best solution against the disease. Oat varieties with various levels of leaf rust resistance for Queensland are available. Many oat varieties that were once resistant have become susceptible to this disease. The leaf rust fungus has the ability to mutate into new and virulent forms which overcome those varieties that were once resistant. Leaf rust is sometimes confused with red-tipped leaf which is not a disease but is believed to be a nutritional disorder.

Infection conditions

— The fungal disease causing this condition survives between cropping seasons on wild oats and volunteer oat plants, which provide a continual supply of spores or inoculum for rust outbreaks each year. The orange spores, seen as pustules on the leaf and leaf sheath, are spread by air.

— Given suitable conditions, it takes one to two weeks for a rust spore to infect and produce more spores on the leaf surface.

— In spring and summer the conditions for rust development are the most favourable with rapid build-up of spores producing serious consequences. In severe outbreak of leaf rust, total loss of the crop can occur.

— Leaf rust can also occur in late autumn in early sown crops (such as February and early March plantings). Early planted crops are usually the worst affected with the epidemic developing rapidly during autumn/early winter.

— Leaf rust is more frequently a problem in warm humid regions such as eastern areas of southern Queensland.

Managing leaf rust

— The best solution to leaf rust is to grow resistant varieties.

— Sowing too early (before mid-March) or too late (after June) will increase the risk of young crops becoming infected and provide a new source of inoculum. This cycle can be broken. For instance, by delaying sowing until after mid-March exposure to leaf rust is reduced early in the season and crops are still able to attain adequate growth before winter.

— Losses from leaf rust can also be reduced by grazing or cutting rust infested crops before the disease becomes severe. For instance, if leaf rust is obvious below the top two leaves on each stem, crops should be grazed or cut regardless of growth stage.

— Plant in wider rows to produce an open canopy, and reduce losses from trampling.

— Maintaining good soil and crop nutrition with nitrogen will minimise the affects of leaf rust.

— There are no fungicides registered for control of rust in oats in Queensland.

Stem Rust

Stem rust is caused by a fungus (*Puccinia graminis avenae*) and appears as dark, reddish brown pustules on leaves, leaf sheaths, stems and heads of the plant. The pustules produced are often surrounded by ruptured leaf and stem surfaces. Stem rust can appear early in the season, but tends to be damaging towards spring and summer. It is favoured by warm, moist conditions typical of late spring and early summer. Normally stem rust is widespread when the crop comes to head.

The disease spreads as spores by wind. As in leaf rust, the fungus survives between cropping seasons in wild oats and volunteer oat plants providing a source of inoculum for the new crops. Stem rust is not considered a problem for grazing oats as it occurs late in the growing season. Stem rust limits oat grain production in Queensland and can cause total loss of the grain. In mild cases the disease produces small and light grains or badly lodged crops. The solution to stem rust in oats is to develop resistant varieties through breeding.

Leaf Spot

Leaf spot (*Drechslera avenae*) appears as small light reddish-brown spots on the leaves and leaf sheaths. It is more common in seedlings and early stages of growth of the oat crop. These spots soon develop into blotches that are oblong to linear with irregular margins. At a later stage, they turn to reddish-brown in colour. In extreme cases, the infected leaf becomes yellow and dries out.

The fungus spreads as spores from plant debris. Infection on plant parts is commonly associated with splashes of rain drops. Spores are often lodged in the hulls of the grain, providing a source for infection.

Leaf spot could be a problem for grazing oats in some years. It frequently appears on the leaves of infected plants early in the season, although with regrowth its incidence could diminish. With serious infection both the yield and quality of the forage can be affected.

Septoria Blotch

Septoria blotch (*Phaeosphaeria avenaria*) first appears on leaves as small, dark brown spots. At a later stage, these spots develop into larger, circular to oval-shaped blotches, surrounded by light yellow areas. These blotches often join together, producing larger lesions. In extreme cases, the infection can kill the entire leaf, if not a larger part of the leaf. The infection may spread to the leaf sheaths and stems resulting in lodging. Septoria blotch appears early in the growing season (i.e. autumn) and spreads rapidly in cool, rainy weather. The fungus is transmitted by spores commonly found in plant debris. These spores spread to young seedling through splashes of rain drops. Septoria blotch is not considered a problem for grazing oats. Although this disease shows up early in the growing season, regrowth is generally free of it. It could be a problem for oats grown for hay or grain.

Bacterial Blight

Bacterial blight (*Pseudomonas syringae*) appears on leaves and leaf sheaths as halo blight and stripe blight. Halo blight initially appears as light green water-soaked oval spots (halos) which eventually turn into brown blotches. Stripe blight appears initially on leaf and leaf sheaths as light green spots. Unlike the spots of halo blight, these spots are not surrounded by a pale green margin. The lesions on the stripe blight eventually lengthen and turn brown, forming water-soaked brown stripes. Blotches produced by both halo and stripe blights often join together to produce large irregular blotches. In both cases, the heavily infected leaves soon turn brown and wither, starting from the leaf tip. At this stage, it is difficult to distinguish the two forms.

The bacteria survive on seeds and plant debris and are often spread by splashes of rain drops, by leaf contact or by insect such as aphids. Bacterial blights could be a problem as it is often widespread early in the season. It could be managed by grazing the crop to remove the infected tissues.

Other Minor Diseases in Oats

Covered smut (*Ustilago hordei*) and loose smut (*Ustilago avenae*): Any seed that may be used for seed crops should be treated with a suitable fungicide before planting.

Barley Yellow Dwarf Virus (BYDV): BYDV is a minor disease in Queensland, and is more important in southern areas of Australia. It is spread by an aphid vector, and sometimes occurs as 'hotspots' in an oat crop. Plants become stunted and leaves become light green in colour with reddish tips often with a mottled appearance. This disease is also confused with red-tipped leaf.

WEED MANAGEMENT

Under favourable growing conditions and regular grazing, chemical weed control is not normally required. Where severe broadleaf weed problem exists, a range of herbicides are available.

Caution is needed when spraying oats as it has much lower tolerance of 2,4-D and MCPA sprays than the other cereals. Check the chemical label before spraying.

Do not under sow medics or vetch with the oats if there is a broadleaf weed problem as they are very sensitive to herbicide spays.

REFERENCES

Janatuinen, E et al., "A Comparison of Diets with and without Oats in Adults with Celiac Disease", *New England Journal of Medicine*, 1995.

Zhou, X., Jellen, E.N., Murphy, J.P., "Progenitor germplasm of domesticated hexaploid oat", *Crop science*, 1999.

6

Local Dispersal and Migration of Insects Pests

The importance of dispersal in insect population dynamics has been largely ignored when formulating integrated pest management (IPM) programs. This is because studies on dispersal by insect pests have focused largely on long-range migration by moths and locusts at high altitudes. For example, the cases of forecasted dispersal reviewed by Pedgley all involve migration by insects over considerable distances. Migration on a smaller scale has been somewhat ignored. Some feel that long-range dispersal, which we conveniently define as less than 5.0 km, may be rare, and more emphasis should be placed on short-range dispersal. Short-range migration is certainly as important biologically as long-range dispersal, and is probably more important agronomically. Laboratory and field experiments concerning migration of agronomic pests on a local scale have largely focused on homopteran insects (Homoptera: Aleyrodidae and Aphididae). Examples are discussed below.

There is no lack of understanding among many academics and growers that migration by pest insects is a phenomenon that impacts crop production. Migration is, however, often omitted as a component of IPM programs at the local level, for two main reasons. To begin, too little is known about the factors that influence migration and dispersal by a particular insect pest. Secondly, testing the impact of this phenomenon on pest populations in agricultural settings is notoriously difficult. This situation is unfortunate, since we require accurate models of insect dispersal to enable realistic forecasting of pest pressure on crops, and this is not possible without detailed knowledge of insect flight behavior and ecology. The interactions between migration and other aspects of pest biology are also critical in developing IPM strategies to deal with mobile pests. For example, in the case of aphids and whiteflies, dispersal of only a few individuals that are insecticide resistant or virus vectors into a crop may give rise to offspring that will cause a significant impact on crop yields.

Dispersal is a term that is inclusive of migratory and trivial flight (insects can disperse by either behavior, or by random locomotion). Dispersal can be accomplished by populations and not individuals. Dispersal may or may not be associated with the characteristics described above. Dispersal is more simply defined as movement that results in an increase in the mean distance between individuals.

Both migration and dispersal may lead to emigration of a pest from one crop and its eventual immigration into another. Local dispersal by insects may be effected by migratory behaviors, or by short host-seeking flights.

LOCAL DISPERSAL BY PEST INSECTS

Aphid

Dispersal by insects on a local scale has been extensively studied in aphids. All of these studies use techniques that show that the insects moved from one place to another, but their behavior during dispersal is not known, i.e., they may or may not be engaged in migratory flight. Thus, we cannot categorically state that the insects were all true migrants in the sense of the definition we have provided. Whatever the behavioral states of the insects, they are primarily relocated to new host plants as a result of being carried by the wind. Some of the aphids that take off and have phototactic flight behavior are evidently carried large distances on the wind. These insects may embark on windborne migratory flights that can carry them for great distances, but the ecological importance of this has recently been challenged. They argue that while long distance dispersal is possible, the majority move locally.

Evidence for local migration by pest insects has been collected from a number of reports on the green peach aphid, *Myzus persicae* (Sulzer). For example, a variant of this species (R2) that is particularly resistant to insecticides was found to be concentrated in northern Britain in samples taken between 1980 and 1984, and the incidence of resistant aphids had not changed appreciably since surveys in 1976. Other resistant variants of this species have been surveyed, and are known to have stable regional distributions. This situation would be predicted by a local dispersal of individuals with limited regional gene flow. In the UK, green peach aphid populations in the south and west are more susceptible to insecticides than those in the north, and thus it seems that there is limited interaction between these populations.

In the early 1960's, potato crops in Idaho, United States (US) were suffering large economic losses to viruses, especially to potato leaf roll virus that is transmitted by green peach aphids and causes stunted growth and yellowing of foliage. A strong correlation was found between the number of aphids per 100 leaves in the potato crop, and the fieldís distance from principal towns. In addition, within field sampling showed a greater number of aphids, and higher disease incidence on the side of the field closest to the towns. Viruliferous aphids and potatoes were found in the towns' gardens. Problems were related to the fact that is was common practice for gardeners to use small, diseased potatoes as seed pieces.

Short duration flights of aphids from the gardens to the surrounding fields were implicated as the cause of the disease outbreaks. Based on this implied local migration by green peach aphids, a control program was implemented in an area-wide program in southeastern Idaho beginning in 1961. There were three main components to this IPM program: 1: supplying seed potatoes certified as being virus free to home gardeners, 2: malathion application to transplants and gardeners' potatoes and 3: removal or treatment of peach and apricot seedlings with oxydemeton-methyl where the insects overwinter. Implementation of this program drastically reduced aphid numbers in the potato crop to less than 1% of the pre-program level, and significantly reduced the incidence of the virus.

Whitefly

This insect has been observed in the US since the early part of this century. It has the ability to transmit a number of viral pathogens and additionally reduces yield through direct feeding damage. More recently, the problems associated with honeydew (whitefly excreta) and sooty mold, *Capnodium* spp., (i.e., stickiness and staining) have become more important for growers across the southern tier of the US and around the world.

Although sampling plans and action thresholds are now being developed for melons, and cotton, *Gossypium* spp., only a few current management programs incorporate more than a few of the strategies we associate with IPM. Growers of most crops are primarily left with one principal option, i.e., the direct application of pesticides. The situation mandates the need for alternative control strategies. We are convinced that a better understanding of flight behavior by *Bemisia* will lead to the development of new cultural control techniques involving computer simulation models that can predict when, where and how far *Bemisia* is going to disperse.

Initial experiments

An early concern, which was pivotal, was whether *B. tabaci* was capable of exhibiting migratory behavior. We had discovered that wingbeat frequency was not significantly correlated with wing loading for these small insects. Larger insects seem to employ a strategy similar to other flying animals, i.e., they compensate for high wing loading with higher wingbeat frequencies. The lack of correlation for these two values in whiteflies led to erroneous assumptions concerning the ability of whiteflies to sustain flight. On the other hand, we knew that the internal lipids in adult females of *B. tabaci* represented 40% of their total dry weight. These compounds are commonly thought to serve as an energy source for insects during flight. For comparison, in the pea aphid, *Acrythosiphum pisum* (Harris), which is known for its migratory ability, lipids account for only 10% of the dry weight. These findings suggested that, despite their small size, *B. tabaci* might possess enough energy stores to engage in extended flight.

Laboratory Behavioral Studies

To investigate a series of hypotheses, a vertical flight chamber was used to examine untethered flight of *B. tabaci.* Blackmer and Byrne characterised the periodicity, intensity and duration of whitefly flight. Behaviorally they reaffirmed that there was a dichotomy between groups of whiteflies relative to flight behavior. Some flew for more than 3 h, others flew sporadically, some flew not at all.

Blackmer and Byrne also examined how whitefly age and host quality (vegetative vs. senescing) affected flight activity. The propensity to takeoff, proportion exhibiting phototactic orientation, and flight duration varied with whitefly age. Longer-duration flights occurred between three and five days of age. Six days following eclosion little flight took place because, among other things, indirect flight muscles and mitochondria in the thorax had started to deteriorate. Host quality also influenced the timing of flight behavior. Whiteflies reared on senescing plants exhibited greater takeoff rates and initiated longer phototactic flights than individuals reared on vegetative plants.

Laboratory Physiological Studies

It is often assumed that there is a trade-off between migratory capability and reproduction. In migrating individuals, nutritional and energy resources are diverted away from growth and reproduction towards building and maintaining flight muscles and energy resources needed for flight. This allocation of resources into either migration or reproduction has been called the "oogenesis-flight syndrome", and has been well documented in alate aphids. We found small, but significant increases, in egg-protein levels were found with longer flights and increasing weights. This was in contrast to the expected.

Lipids are the most suitable materials for storage of energy reserves. Compared to carbohydrates, lipids can supply as much as eight times more energy per unit weight. In addition, twice as much metabolic water is produced by metabolising lipids instead of carbohydrates. A high lipid content associated with migration has been found in many aphids and some bark beetles. In the sweet potato whitefly we found no evidence that lipids are stored for dispersal of less than 3 km. However, flight chamber experiments indicate that in females, lipid content is correlated with flight duration. Thus, lipid reserves could be of more importance for long distance migration.

Understanding the factors leading to production of migratory morphs is important for IPM strategies. Physiological studies can lead to valuable insight into migratory capabilities and will allow us to estimate the effective range of dispersal of insect pests.

Field studies with whiteflies

In order to corroborate laboratory observations, field experiments were conducted beginning in 1992. Initial experiments concentrated on the impact of wind flow on the directionality of flight by *B. tabaci.* A 0.73 ha circular field was planted with cantaloupe, *Cucumis melo* L. The area around the field was surveyed for placement of muffin fan traps along a series of 16 transects placed 22.5^0 apart. These traps are described in Byrne et al., They allowed the capture of living whiteflies.

The evening before each sampling day the melon field was dusted with fluorescent dust. Tracking insects from where they alight to their landing place is possible in the field, and has been done with small insects with some success using the techniques listed separately. The traps were turned on the morning of each sampling day (at approximately 6:00 a.m.). After 3 h, whiteflies marked with dust were collected from the traps. For each sampling date in 1992 the relationship between wind flow and the number of whiteflies captured in traps in the direction opposite the prevailing wind was evaluated. Prevailing wind directions from the northeast and trap catch were in agreement for all dates (when the wind was from the northeast > 90% of the whiteflies were caught in traps placed to the southwest of the marked field). This provided empirical evidence that whitefly migration is wind-aided.

Following 1992, traps were located as part of a grid to the southwest of the marked field. Maps of fields and trap locations were digitised using a digitising tablet. Data were analysed using the Geostatistical Information System (GIS). The principal feature evident when we examined the data from our field experiments was that whitefly distribution following dispersal was patchy. Almost all of the sampling dates show greatest trap catches of marked insects near the marked field and on the periphery of the grid than in the center of the grid. Pooling data from all sampling dates in 1993 produced a line graph indicating a bimodal distribution. It is evident that the trap catches did not indicate a simple diffusion pattern, a pattern that might be expected if dispersal was passive across featureless topography.

This is in contrast to patterns seen by Turchin and Thoeny who described dispersal by southern pine beetles, *Dendroctonus frontalis* Zimm., using a diffusion model. We believe our pattern may be explained by the existence in whitefly populations of two behaviorally distinct types; trivial flying individuals and migratory individuals as described by Byrne and Houck and Blackmer and Byrne. Insects caught in the immediate vicinity of the source field are likely to be trivial fliers, while those caught on the grid perimeter are thought to be migratory individuals. The explanation for the relative paucity of whiteflies caught in the middle of the grid is that migratory whiteflies do not fly towards green vegetation until they have flown for a given period of time.

In our studies, the greatest distance marked whiteflies have been trapped is 2.7 km from a source field. Since whiteflies were consistently captured in the most distant traps, the effective migrational range of *B. tabaci* may be considerably greater than this. Future experiments will be conducted with an expanded grid to further define the effective migrational range (i.e., the distance they can travel and still initiate a colony) of *B. tabaci*.

Field research is still ongoing, and these experiments have already revealed a few surprises. For example, we have found that whiteflies were caught in traps placed at the edge of the source field, at over 20 ft above the crop. Traps were in place just before dawn, because the air starts to move at sunrise, and flight is observed as soon as there is sufficient light available to stimulate take-off. Thus, it would seem that the migrants are taking maximum advantage of the low wind speeds to gain height and take them away from the plant boundary layer.This has far-reaching implications for our understanding of whitefly migration and their dispersal on the wind. That whiteflies can take off and ignore plant cues in the field also reinforces our faith that the behaviors observed in the laboratory are those employed by the insects in the field.

Development of predictive models

Ideally, our knowledge of insect migration would lead to the development of simulation models, that serve as an aid to growers in making management decisions. To our knowledge, no such models exist as of yet. This is, in our opinion, partially due to some inherent difficulties in developing and implementing such models.

The "art" of modeling involves reducing a complex biological reality down to a few mathematical statements. For example, one could treat a field of melons, as one unit, with specified size, location, and time of planting. However, the development of whiteflies in a given melon field and their migratory tendencies, are likely to be influenced by the many other factors, like watering regime, amount of fertiliser applied, soil type, or disease incidence. In order to incorporate these factors into the model we need to know the effect of each of these factors on whitefly biology. If one of these factors turned out to be critical for the model, growers would have to spend time (and money) collecting data on this variable. That would severely diminish the value of these models as a management tool.

In addition, one typically has only data available from a small number of successive years for the development of migration models. Thus, the model will be based on a subset of potential environmental conditions. For example, the years for which data were available could be characterised by relatively cool summers. Whether the model thus developed could be extended to years with different temperature regimes can only be examined by collecting data over longer time periods.

ENVIRONMENTAL RISK AND PEST MANAGEMENT

Considering questions of environmental risk as we take action against pests has become an increasingly important issue. Environmental risk involves technical understandings of risk, public perceptions, and public influence, as well as fundamental questions about the principles and goals for pest management. Although some questions of risk date back to the origins of the concept of pest management, many modern issues are relatively recent developments.

A key question to be asked is if pest management even needs to address issues of environmental risk. As Perkins has pointed out in his seminal book, Insects, Experts, and the Insecticide Crisis, the goal of many workers in pest management is to address the needs of farmers. As we consider issues in environmental risk, we must recognise that there can be conflicts between the needs of agricultural producers and the concerns of society in general. Is it part of the explicit goals of pest management to deal with these greater societal concerns on issues like environmental risk? In other words, is pest management only about dealing with pests and maintaining profitability or are other, noneconomic considerations important?

The goals of pest management include economic sustainability through minimising the economic impact of pests, ecological sustainability through employing management tactics to minimise selection pressure, and environmental sustainability impact of management tactics on the environment. It is this latter point, of course, that is the focus of these comments, and it seems an important one to make about any technology.

While society may be willing to accept some environmental deterioration in association with certain technologies (air pollution associated with generating power and use of automobiles is an obvious example), nevertheless, we would like to have technologies with minimal impacts. Because pest management is a philosophy addressing how we should employ technologies against pests, it follows that minimising environmental impacts of those methods should be one of our goals.

Pimentel and Lehmen offer various perspectives on issues of pesticides and environmental quality. Regarding pest management broadly, Pedigo and Higley and Higley and Pedigo offer perspectives on improving responsiveness to environmental questions through consideration of the economic injury level.

Risk and Risk Perception

Public perceptions of pesticide safety increasingly influence pesticide legislation and regulation. Unfortunately, although public concerns about pesticides do, and should, contribute to the formation of pesticide legislation and policies, perceived risks may contribute to irrational, even counterproductive, regulations. Misunderstandings about public perceptions of risk and failure to address the key issues of risk perception, are important barriers to consensus about the risks posed by pesticide use.

Although definitions of risk vary, most recognise risk as:

$$R = P * C \tag{1}$$

where R is risk, P represents the chance or probability of an undesirable event, and C represents the adverse consequences of the event. You might say that risk is a measure of how often something will happen and how bad it will be. Differences in risk perception involve different perspectives on both probability and consequence.

Public Perception

Risk perception research began in earnest soon after the accident at the Three Mile Island Nuclear Power Plant in Harrisburg, PA. Researchers were intrigued that expert estimates of risk from technologies such as nuclear power did not mirror public estimates of risk from the same technologies. Therefore, risk perception research seeks to understand how people perceive risk presented by different circumstances and activities.

The public continually ranks certain risks much higher than experts. According to Slovic, "Lay people can assess annual fatalities [from a given factor] if they are asked to (and produce estimates somewhat like the technical estimates). However, their judgments of `risk' are related more to other hazard characteristics (for example, catastrophic potential, threat to future generations) and, as a result, tend to differ from their own (and experts') estimates of annual fatalities."

Perceptions of Pesticide Risk

Public perceptions of risk from pesticides usually are greater than the risks determined by experimentation. Experts determine risk from pesticides based on empirical estimates of chronic and acute human and animal toxicity and environmental fate. Although in most cases human health risks associated with pesticide use are very low, the public consistently ranks pesticide use as being very risky. Indeed, the public perception of risk posed by most synthetic chemicals is inordinately great. Too often, people who work with pesticides and have a more complete understanding of the risks often dismiss public concerns as irrational or misinformed. Consequently, many technical people react defensively to public concerns. Even worse, some regard public fears as so irrational that they do not attempt to explain risks and the risk assessment process to lay people. This is dangerous behavior, because the public needs to understand and be involved in pesticide issues to improve the policy making process.

Responses to public concerns by experts are often inadequate, doing little to seriously address public risk perceptions. A traditional approach has been to respond to public concerns by comparing highly visible, well-understood risks to less visible, less well-understood risks. Some responses have compared the risks from pesticide exposure to the risks presented by everyday activities, such as driving an automobile or riding a bicycle. Although most risk comparison statements by pesticide experts are more accurate than public perception of the risks, these responses do little to correct the public's misconceptions about the unfamiliar risks of pesticide use. Most experts fail to consider the criteria the public actually use to evaluate risks.

Risk Criteria

Although the risks associated with driving an automobile are more serious, the public perceives using pesticides as being much riskier. Researchers recently have explored the mental processes people use to assess risks from many types of modern technologies and activities. This body of research demonstrates that lay people use a different set of criteria than experts for evaluating risk. Specific factors that influence public risk perception include:

- Control - the ability of the individual or society to control the risk.
- Catastrophic potential - the possibility of fatalities or ill effects grouped in time and space as an epidemic.
- Dread - the fear of the possibility of serious delayed effects, such as cancer. Dread is related to catastrophic potential, but the impact does not necessarily need to be grouped in time or space.

— Familiarity - the degree of familiarity lay people have with risk.
— Equity - the equal distribution of risks and benefits throughout society.
— Level of knowledge - the general understanding lay people have with the process or activity posing the risk.
— Voluntariness of exposure.
— Effects on children and future generations - concerns about possible delayed effects on humans and the environment posed by the risk.
— Clarity of benefits - represents the awareness and understanding of the benefits provided by the activity posing the risk.
— Media attention.
— Trust in organisations or institutions.

The public uses these characteristics to judge the acceptability of a risk, rather than using risk estimates based on experiments. Such a view of risk is more encompassing and powerful than simple estimates of mortality. In addressing risk, we should especially consider catastrophic potential, control, level of knowledge, and effects on children and future generations.

One reason the public may view pesticides in food as riskier to their health than the natural carcinogens in common foods and beverages is because consumers have no control over the pesticide content of food. In contrast, people can choose which types of foods they eat, and, therefore, they have a sense of control over this activity. Similarly, people may accept the risks from driving an automobile much more readily than the risks from pesticide exposure, because they have control over the automobile.

The public also will not readily accept the risk associated with a potential catastrophe. And yet, catastrophic potential is seldom discussed by the experts.

For example, public fear of nuclear power is largely based on vivid images of millions of radiation deaths, massive environmental devastation, and threats to future generations should an accident occur. Consequently, the public's perception of risk from nuclear power is far greater than the experts'. More to the point, the pesticide disaster at Bhopal, India, although it was an industrial accident, is a vivid example of the catastrophic potential of pesticides.

The public views pesticides as having both catastrophic potential and posing a dread risk. First, pesticides are feared because of their possible harmful delayed effects such as cancer. Second, the public is greatly concerned about environmental damage caused by pesticide use. Widespread knowledge of the problems with biomagnification and contamination of groundwater by pesticides and their metabolites, coupled with the reality that pesticides are among the most extensively broadcast synthetic chemicals in the environment, makes dread risk a very real concern to the public.

Coverage of a risk by the mass media can reinforce existing public risk perceptions. Media coverage often reveals the dangerous activities that are so pervasive in today's technological world. The mass media provide visual images of risks, enhance the imaginability of a problem, and concentrate on the human aspects. The media may also remind the public that they have little or no control over these activities.

The public is unlikely to accept risks from an activity if experts openly disagree over the magnitude and nature of the risk. Indeed, the public knows that experts have advocated the use of pesticides without full knowledge of their potential hazards, DDT being a prime example. The public views pesticides as risky, in part, because of a justified lack of confidence in expert opinion. Once the credibility of experts and government institutions has been damaged, it is very difficult to regain public confidence. To complicate matters further, the public realises that experts can be found to support any position in a risk argument because attitudes about risk are inherently subjective.

Public perception of risk also is influenced by individual and group goals, values, and politics. Some groups may want to see all pesticides banned; other groups may want less government regulation of pesticides. The goals and values expressed by individuals in these diverse groups will influence their risk perceptions. Similarly, members of society do not equally accept the benefits from a technology.

The understanding, or clarity, of benefits associated with pesticide use also is considered by the public in conjunction with pesticide risks. Where benefits relate directly to questions of risk, for example, accepting one risk (such as pesticides) to avoid a greater risk (such as tick- or mosquito-borne diseases), benefits may have a direct bearing on risk perception. More commonly, discussions of benefits focus on economic impacts or crop yields which the public does not regard to be as important as health or environmental risks. Assessing benefits is very much a matter of opinion and because benefits do not directly apply across society, discussions of pesticide benefits are unlikely to quell public concerns about pesticide risks.

RISK IN PEST MANAGEMENT

Tactics and Frequency of Use

Historically, two approaches have dominated the question of environmental quality in pest management: frequency of use and choice of management tactic. Regarding frequency of use, many efforts in pest management are directed at reducing unnecessary management action (typically, eliminating unneeded pesticide use). The development of the economic injury level (EIL) and associated economic thresholds (ET) is one clear reflection of this trend.

In most cases where the issue has been studied, the use of pesticides is reduced with thresholds. These reductions are a consequence of conservative attitudes toward protecting crops with pesticides. Relative to other costs of production, pesticides often provide relatively inexpensive insurance against catastrophic losses from pests. However, although thresholds have been important in reducing frequency of use, it is not an intrinsic feature of EILs or ETs that necessarily lead to a reduction in use. For example, in situations where a crop is not intensively managed, such as with forages or other low-value crops, the use of thresholds can cause an increase in pesticide use.

The second important way that pest management can address environmental risk is through the choice of management tactic. If we associate most risk with the use of pesticides, then the use of alternative tactics such as host plant resistance, cultural techniques, or biological control, provide more environmentally-friendly alternatives. Sometimes we will use such tactics in combination with pesticides, in which case the use of alternative tactics may impact frequency of pesticide use.

Although alternative practices are usually regarded as less risky than pesticides, risk is associated even with these tactics. For example, an increased chance of soil erosion is associated with certain rotational patterns or tillage practices. Similarly, very high levels of an antibiotic factor in a crop can, of themselves, prove to be of risk. Even with biological control, there have been recent arguments about assessing risk to native species from natural enemy introductions. One problem with the use of alternative tactics is that most of these tactics have to be used in a preventive fashion. One of the outstanding limitations in pest management, particularly relative to finding alternative practices, is that we have few alternatives to pesticides for therapeutic management actions.

One development that does relate to the choice of management tactic for improving environmental quality has been changes in pesticide development over the past thirty years. Since the discovery of the deleterious environmental impacts of DDT and other chlorinated hydrocarbons, there has been a very strong effort to find compounds that not only have good properties with regard to human safety, but also good properties with regard to environmental safety. Environmental safety includes such considerations as effect on non-target species (birds, mammals, fish, etc.) and reduced potential for long-term environmental contamination. In this arena, there have been remarkable improvements in environmental safety.

Development of the pyrethroids and some of the new, so-called "biorational compounds" clearly have resulted in chemicals that present much less environmental risk. Ironically, in developing such compounds a conflict arises between desirable properties in terms of affecting pests (such as longevity) versus desirable properties from an environmental standpoint (such as short life in the environment). Nevertheless, the gradual transition in insecticide development toward environmentally safer compounds is a major contribution to risk reduction.

Environmental Risk and the EIL

Pedigo and Higley have argued that by considering components of the EIL from an environmental standpoint, it may be possible to identify approaches that could be pursued in pest management to try to improve environmental quality. In their original article and a subsequent paper, they discuss potential approaches. Preferential taxation of pesticides based on environmental risk is one approach they identified, although they note that this would undoubtedly have the potential to reduce proof of producer profitability and therefore runs in conflict to other goals of pest management.

Regarding plant response to pest attack, they note that emphasis on improving the ability of plants to tolerate or compensate for pest injury offers the potential for reducing the need for intervention and for providing longer-term solutions to pest management tactics. In reviewing injury per pest, they note that although this has been proposed with respect to medical pests, there seems to be little likelihood that it would be of use for agronomic pests, given that changing the amount of pest attack requires changes in an entire pest population. However, one example that may be useful is with respect to weeds, where there are demonstrations that extremely low concentrations of herbicides can be used to impair the competitiveness of weed species sufficiently to reduce the need for subsequent management action.

A final area they identify has to do with the proportion of the pest population affected by management action. A number of studies suggest that current rates of pesticide use are well in excess of what is actually needed for reliable management, so reduced pesticide rates may be a simple approach that could considerably reduce pesticide loads, provided that pesticides are properly applied.

Environmental EIL

The Pedigo and Higley arguments regarding environmental quality and the EIL are largely theoretical, in that they have tried to view the EIL as a framework for identifying areas in pest management where efforts might be directly focused on questions of environmental quality. Another, more controversial, approach is to use the EIL itself to address environmental risk directly. Higley and Wintersteen first proposed this approach, which has a relatively simple premise. Because the EIL relates insect damage to economic costs of that damage, why not include in the economic costs the indirect effect of the pest on environmental quality. In other words, in addition to assigning a cost to crop losses as part of the EIL, why not also assign an environmental cost, which is a reflection of the environmental risk posed by the management tactic.

The central difficulty in such a proposal is establishing exactly what environmental costs would be. As one solution, Higley and Wintersteen proposed a scheme for developing estimates of the environmental costs for different pesticides. Once they have these cost estimates, it is relatively straightforward to reflect costs in an EIL. The procedure Higley and Wintersteen proposed employs a technique called contingent valuation, which has been used by economists to try to put a dollar value on what are called "non-market goods" .

Contingent valuation uses opinion surveys to try to estimate costs. For example, a typical use of a contingent valuation survey might be to ask how much you would be willing to see your taxes increase to clean up a polluted lake. Higley and Wintersteen used contingent valuation to ask field crop producers in the Midwest how much they would be willing to pay to avoid different levels of environmental risk. From these cost estimates, they associated a value with different levels of risk. They went on to define risk by looking at chemical properties of pesticides and the effect of pesticides on a variety of non-target species. Additionally, as part of their survey procedure they also asked field crop producers to define the relative importance of these different areas of environmental risk.

REFERENCES

Smith, R. D., "Background, use, and beneficts of blast freezers in the prevention and extermination of insects", en *Biodeterioration 6: papers presented at 6th Interational Biodeterioration Symposium, Washington DC, August 1984.*

Stranger, Carolla; Brandis, Leanne, "Insect pests and their erradication (non-chemical methods)", en *Australian Library Journal,* 1992.

7

Plant Resistance to Insects

The role of plant resistance as a direct control tactic in insect integrated pest management (IPM) will be emphasised. The use of insect-resistant varieties as a component of an IPM strategy will be provided.

Ecological results of agriculture: In contrast to natural ecosystems, most agricultural crop production systems are ecologically unstable, non-sustainable, and energy dependent. Man has through plant domestication and cultivation practices interfered in many ways with species diversity and natural defense mechanisms of plants. Cultivated crops originated from genetically diverse plant types. However, crop plants are now grown in large, genetically homogeneous stands, a practice that decreases genetic and species diversity and increases the likelihood of economically significant insect pest infestations. Defense mechanisms of plants are re-created in resistant plants.

Plant defense mechanisms include escape in space and time, incompatible biological associations, physically and chemically derived barriers, and accommodation by replacement or repair of damaged plant parts. By re-creating plant defenses, genetic resistance to insect pests plays, in an environmentally compatible manner, a vital role in the attempt to enhance ecological stability in agricultural crops.

An IPM direct control tactic: Plant resistance to insects is one of several cultural control methods. Cultural control methods involve use of agronomic practices to reduce insect pest abundance and damage below that which would have occurred if the practice had not been used. In IPM, plant resistance to insects refers to the use of resistant crop varieties to suppress insect pest damage. Plant resistance is intended to be used in conjunction with other direct control tactics.

Insect-resistant plant: Definitions of an insect-resistant plant are many and varied. In the broadest sense, plant resistance is defined as "the consequence of heritable plant qualities that result in a plant being relatively less damaged than a plant without the qualities." In practical agricultural terms, an insect-resistant crop cultivar is one that yields more than a susceptible cultivar when confronted with insect pest invasion. Resistance of plants is relative and is based on comparison with plants lacking the resistance characters, i.e., susceptible plants.

Effect of insect pest-plant host relationship: Insect-resistant crop varieties suppress insect pest abundance or elevate the damage tolerance level of the plants. In other words, insect-resistant plants alter the relationship an insect pest has with its plant host. How the relationship between the insect and plant is affected depends on the kind of resistance, e.g. antibiosis, antixenosis (non-preference), or tolerance.

Antibiosis resistance affects the biology of the insect so pest abundance and subsequent damage is reduced compared to that which would have occurred if the insect was on a susceptible crop variety. Antibiosis resistance often results in increased mortality or reduced longevity and reproduction of the insect.

Antixenosis resistance affects the behavior of an insect pest and usually is expressed as non-preference of the insect for a resistant plant compared with a susceptible plant.

Tolerance is resistance in which a plant is able to withstand or recover from damage caused by insect pest abundance equal to that damaging a plant without resistance characters (susceptible). Tolerance is a plant response to an insect pest. Thus, tolerance resistance differs from antibiosis and antixenosis resistance in how it affects the insect-plant relationship. Antibiosis and antixenosis resistance cause an insect response when the insect attempts to use the resistant plant for food, oviposition, or shelter.

Advantages to the use of insect-resistant crop varieties: Use of insect-resistant crop varieties is economically, ecologically, and environmentally advantageous. Economic benefits occur because crop yields are saved from loss to insect pests and money is saved by not applying insecticides that would have been applied to susceptible varieties. In most cases, seed of insect-resistant cultivars costs no more, or little more, than for susceptible cultivars. Ecological and environmental benefits arise from increases in species diversity in the agroecosystem, in part because of reduced use of insecticides. Increases in species diversity increase ecosystem stability which promotes a more sustainable system far less polluted and detrimental to natural resources.

The IPM concept stresses the need to use multiple tactics to maintain insect pest abundance and damage below levels of economic significance. Thus, a major advantage to the use of insect-resistant crop varieties as a component of IPM arises from the ecological compatibility and compatibility with other direct control tactics. Insect-resistant cultivars synergise the effects of natural, biological, and cultural insect pest-suppression tactics. The "built-in" protection of resistant plants from insect pests functions at a very basic level, disrupting the normal association of the insect pest with its host plant. The compatible, complementary role plant resistance to insect pests plays with other direct control tactics is, in theory and practice, in concert with the objectives of IPM. All crop cultivars should contain resistance to insect pests.

Plant resistance to insect pests has advantages over other direct control tactics. For example, plant resistance to insects is compatible with insecticide use, while biological control is not. Plant resistance to insects is not density dependent, whereas biological control is. Plant resistance is specific, only affecting the target pest. Often effects of use of insect-resistant cultivars are cumulative over time. Usually the effectiveness of resistant cultivars is long-lasting.

The role of plant resistance to insects in IPM has been well defined, at least in theory. However, the specific role a resistant cultivar plays in a particular IPM situation is crucial to successful deployment of the resistant cultivar. The impact of the resistant cultivar on standard cultural, biological, and insecticidal control methods should be well defined. Likewise, the impact of each of these control tactics on the resistant cultivar also must be defined.

Several definitions have been used to convey the relative level of resistance in a plant. However, the problem of quantifying resistance continues to be a problem influencing farmer acceptance of insect-resistant cultivars. A better way to define resistance levels in agronomically improved resistant cultivars is through quantified comparisons of insect pest damage or plant yield loss of susceptible cultivars. Once insect pest abundance or damage to yield-loss relationships have been determined, economic threshold levels can be determined and combined with factors such as crop value and insect pest control costs to develop dynamic thresholds for use by producers.

Dynamic thresholds provide a description of resistance and can reduce crop loss risk because limitations are known and remedial action can be taken when necessary. By using this system to define relative differences in insect pest resistance between cultivars, it may be possible to simply indicate that a resistant cultivar has a higher economic threshold level than a traditional susceptible cultivar.

Sorghum midge in sorghum: The sorghum midge, *Stenodiplosis sorghicola* (Coquillett), is a ubiquitous and serious insect pest of sorghum, Sorghum bicolor (L.) Moench. It is a key insect pest of sorghum throughout the southern production region of the United States. Each ovipositing female sorghum midge lays about 50 eggs between the glumes of flowering spikelets of cultivated sorghum or johnsongrass, Sorghum halepense (L.) Pers. Larval feeding on the developing ovary inside the spikelet prevents normal kernel formation and results in direct grain loss. A single larva is sufficient to destroy a kernel; the entire crop can be destroyed.

A sorghum midge life cycle is completed in 16-18 days under favorable temperatures. This relatively short generation time allows for multiple generations of the insect each season. Sorghum midge abundance can increase each generation when flowering sorghum hosts are available and weather conditions are favorable. Sorghum midges can become especially abundant when successive plantings of the crop allow the flowering period to be extended in a locale.

Johnsongrass is essential for survival of the sorghum midge. In the spring, sorghum midges in an area emerge from overwintering before cultivated sorghum is flowering and lay eggs in flowering spikelets of johnsongrass. Typically, a second generation is produced in johnsongrass before cultivated sorghum in an area is flowering. Sorghum midges disperse to the first flowering fields of cultivated sorghum, and within a single additional generation usually have increased enough in abundance to cause economic damage to sorghum.

Management tactics used to suppress abundance of sorghum midge include pest avoidance by planting sorghum early and uniformly in an area, destruction of johnsongrass, abiotic and biotic natural control involving indigenous parasites and predators, insecticide use, and use of resistant sorghum varieties. The major management tactic is early and uniform planting of sorghum in an area so that sorghum flowers before sorghum midge abundance is high enough to cause economic damage. The nature of sorghum midge biology and characteristics of sorghum panicle flowering make insecticide use less effective and more expensive than desirable.

Significant progress has been made in developing sorghum inbred lines that, under high levels of sorghum midge infestation, produce sorghum midge-resistant hybrids with excellent grain yield. This management tactic has received much attention during the past decade. The major resistance mechanism of resistant varieties is non-preference for or interference with adult oviposition caused by differences in spikelet flowering time. Spikelets of resistant sorghum flower during early morning hours before adult female sorghum midges occur in the field. Spikelets of susceptible sorghums flower later in the morning when sorghum midges are present in the field. The consequence of the resistance mechanism is that sorghum midge-resistant hybrids are at least five times less damaged than susceptible hybrids at the same level of sorghum midge abundance.

Consequently, the economic threshold level for sorghum midge-resistant hybrids is five times higher than for susceptible hybrids (Table 1). The elevated economic threshold significantly enhances the effectiveness of other management tactics. In many cases, the resistance level is high enough to prevent economic damage and eliminate the need for insecticides. Also, when sorghum midges become very abundant, insecticide use requirements are much less, and the application(s) is more effective than for sorghum midges infesting susceptible sorghum hybrids.

Resistance is not intended to always provide the sole means of control but is used as a direct control tactic that, in combination with others, provides desired protection of the crop from the insect pest.

Table 1. Economic injury levels for sorghum hybrids infested by sorghum midges.

Economic injury level for sorghum midge-susceptible hybrids											
Per acre control cost ($)	*Crop market value ($) per acre*										
	100	120	140	160	180	200	220	240	260	280	300
3	1.2	1.0	0.9	0.8	0.7	0.6	0.6	0.5	0.5	0.4	0.4
4	1.6	1.3	1.1	1.0	0.9	0.8	0.7	0.7	0.6	0.6	0.5
5	2.0	1.7	1.4	1.3	1.1	1.0	0.9	0.8	0.8	0.7	0.7
6	2.4	2.0	1.8	1.5	1.3	1.2	1.1	1.0	0.9	0.9	0.8
7	2.7	2.3	2.0	1.8	1.6	1.4	1.3	1.2	1.1	1.0	0.9
8	3.0	2.7	2.3	2.0	1.8	1.6	1.5	1.3	1.2	1.1	1.1
Economic injury levels for sorghum midge-resistant hybrids											
3	6	5	5	4	4	3	3	3	3	2	2
4	8	7	6	5	5	4	4	4	3	3	3
5	10	9	7	7	6	5	5	4	4	4	4
6	12	10	9	8	7	6	6	5	5	5	4
7	14	12	10	9	8	7	7	6	6	5	5
8	15	14	12	10	9	8	8	7	6	6	6

Greenbug, Schizaphis graminum (Rondani), in sorghum: This aphid has been an insect pest of small grains in the United States for almost 100 years and a key insect pest of sorghum since the appearance of biotype C greenbugs in 1968. Management approaches to dealing with biotype C after 1968 included using extremely low dosage rates of registered insecticides. Treatment need was based on knowledge of the insect's seasonal abundance trends and economic threshold levels. This IPM approach reduced greenbug abundance below numbers capable of causing economic yield loss and preserved natural biological control agents that prevented greenbug resurgence and secondary pest outbreaks. However, failure of sorghum growers to adopt this IPM strategy and the continued use of high rates of organophosphorus insecticides resulted in high levels of insecticide resistance. Another management tactic was needed.

Development of greenbug-resistant sorghums was initiated in 1969. As a result, greenbug-resistant hybrids were commercially available in 1976. Tolerance is the primary resistance mechanism of greenbug-resistant sorghums. Tolerance is ecologically advantageous because natural enemies are maintained that complement the resistance and suppress greenbug abundance. Also, the incidence of secondary pest outbreaks is reduced.

Unfortunately, numerous greenbug biotypes have occurred. As used here, biotype refers to individuals or populations that differ from others based on criteria other than morphology, such as parasitic (virulence) ability. Greenbug biotypes are characterised on the basis of differential host plant response within the greenbug species. Currently, 11 biotypes of greenbug have been identified although only biotypes C, E, I, and K are virulent on sorghum in nature. The importance of biotype K is yet to be determined. Despite the consistent occurrence of greenbug biotypes, greenbug-resistant sorghums have been a major component of sorghum IPM.

Future possibilities: Much interest in biotechnology relative to developing insect-resistant plants is in methods known collectively as genetic transformation, rDNA methods, or genetic engineering. These methods enable transfer of a resistance gene that could not be transferred by traditional sexual hybridisation. As important, however, are evolving molecular genetic techniques that are providing exciting opportunities to better understand the genetic relationship between an insect and a resistant plant.

DNA marker technology is currently used to map and identify specific gene structures conferring resistance traits in plants. Understanding the genetics of resistance in plants will provide the knowledge to improve resistance deployment strategies. However, it is becoming evident with regard to plant resistance to insects that it is equally, if not more, important to understand the genetics of the insect attacking the plant. However, regardless of the technology used to develop insect-resistant plants, the technology will have to be used as a component of IPM in the same way traditionally developed insect-resistant plants are used today.

As new technology has begun to be used, there is increasing concern over the likelihood of formation of insect pest biotypes that can overcome resistance. Many factors are associated with the ability of an insect to overcome plant resistance. A popular theory is that resistant plants exert selective pressure for virulent individuals that become predominant types (biotypes). This theory of biotype selection often is related to theories of evolutionary biology. For example, one evolutionary biology theory is that selection by insects causes plant populations to acquire resistance traits, and selection imposed by these resistance traits gives the insect the ability to exploit these plants. The implication is that insects with a long-term association with a host-plant species have the genetic plasticity to overcome defense strategies of the host plant. That is, resistance in plants that evolved with an insect could be susceptible to being overcome by genetic changes in the insect pest.

Theories on gene conservation and management are major subjects of scientific discussion and experimentation. Strategies for preservation of insect resistance in crop varieties include use of multiple resistance genes conferring different causes of resistance, sequential (time and space) deployment of resistance, and maintenance of refugia to support avirulent individuals to mate with virulent individuals in the insect population. Concern over insect biotypes probably will result in much wiser use of genetic opportunities that now are available or will be available in the future. A major mistake would be to attempt to use gene-altered plants as the sole control tactic for an insect pest rather than a component of IPM. To do so would be to repeat the mistakes of the insecticide era.

HOST PLANT RESISTANCE AND CONSERVATION OF GENETIC DIVERSITY

Urgency of Global Genetic Conservation

The world's biota are under siege. A human population nearing 6 billion places unprecedented pressures on the biosphere.. Approximately 95% of the terrestrial surface is now occupied by human settlements or ecosystems managed for food and materials production. As a result, natural ecosystems are destroyed and fragmented, species are destroyed or doomed, and global genetic diversity is diminished. The scale of this destruction directly threatens human civilisation, which is dependent in numerous ways on biological diversity. The enormity of the crisis has stimulated international efforts to conserve biological diversity and to ensure sustainable use of its components.

Conserving Agricultural Genetic Diversity

Before the extent of this crisis was recognised, genetic diversity in agricultural crops began to erode as political and social changes of the industrial era transformed agriculture. Genetic conservation began early this century when foresighted agriculturists saw that the loss of primitive varieties or "landraces" meant loss of genetic variation essential for sustained crop improvement.

Pioneers like Soviet geneticist Nikolai Valivov and USA scientist Harry Harlan encouraged and conducted systematic collection of the seeds of indigenous crop varieties and developed an understanding of geographic distribution of genetic variation in crops. The extent of this variation is impressive to those of us accustomed to the uniform appearance of crop plants in use today. For example, indigenous cultivars of maize from Mexico can vary in their time to maturity from 60 days to 16 months, the plants can range from 40 to 700 cm in height, and ear lengths range from 4 to 40 cm! Variation in fruit form in tomato (*Lycopersicon esculentum*) and leaf type in *Brassica oleracea acephala* from the Mediterranean are striking morphological manifestations of genetic diversity. Morphological variation is just one indication of variation in countless physiological, chemical, and developmental traits.

In the early 1970s, international concern escalated to alarm over the continuing erosion of crop genetic resources despite efforts to forestall it. The devastating epidemic of southern corn leaf blight (*Helminthosporium maydis*) in the USA in 1970 occurred because 70% of the corn in grown the USA had been developed using a single source of cytoplasm that was susceptible to the pathogen. This and other evidence that genetic uniformity was dangerous triggered a more concerted effort to conserve crop genetic diversity. The The National Plant Germplasm System of the USDA (NPGS) was restructured, and the International Agricultural Research Centers (IARCs) were united under the direction of the Consultative Group on International Agricultural Research (CGIAR) and the International Board of Plant Genetic Resources (IBPGR).

Today germplasm in these collections and those of other private and public institutions around the world contain over 3,600,000 accessions representing at least 100 crop species and their wild relatives. Among the over 200 national collections, the largest is the NPGS with over 500,000 accessions. Another 500,000 accessions are held in collections of the IARCs. These collections consist of propagative tissues, usually seeds, maintained in cold storage and periodically propagated to preserve their viability and genetic identity. This technique is known as *ex situ* conservation. The collections are curated to provide plant breeders and other scientists ready and free access to facilitate crop improvement. The *ex situ* collections are essentially the only genetic resource used for crop improvement.

Pest-Resistant Crops and Genetic Diversity

Traits conferring host plant resistance (HPR) to insects and pathogens are among the most important for crop improvement. There is probably no crop into which pathogen resistance has not been widely introduced using genes obtained from the international *ex situ* germplasm collections. Insect resistance has also been introduced into several hundred crop varieties during the last 20 years and its importance is increasing as insecticides lose efficacy due to pest adaptation or are removed from use to protect the environment and human health. Often, successful crop production is impossible without resistance to insects and pathogens that otherwise cannot be economically controlled. In many cases, multiple genes are required for sustained resistance to counter pest adaptation. Thus, maintaining agricultural productivity to meet world food needs depends on access by agricultural scientists to plentiful sources of HPR genes.

Existing *ex situ* collections may not be adequate for HPR breeding in the future, however. Pest resistance genes are rare and predominantly found in unimproved varieties or wild accessions. For example, resistance to insects is found in only 0.01 to 2% of rice accessions and much of this occurs in exotic land races. In potato, high levels of resistance to the green peach aphid (*Myzus persicae*) has been identified in about 6% of examined accessions of wild *Solanum* species, but in 0% of over 360 accessions of *S. tuberosum* and other cultivated *Solanum* species. The pattern is similar for resistance to other insect pests of potatoes. In cultivated tomato (*Lycopersicon esculentum*) insect resistance is rare, but it is more prevalent in wild accessions of *L. esculentum* var. *cerasiforme* and common in more distantly related *Lycopersicon hirsutum*.

Pathogen resistance too is most frequently found in unimproved or wild accessions. Rice gets resistance to two of Asia's four main rice diseases from a single wild rice accession from central India. Resistance to 12 different diseases in cultivated tomato originates in wild *Lycopersicon* species. Wild crop relatives have yielded pathogen resistance in barley, cassava, sweet potato, sunflower, grapes, tobacco, cacao, sugarcane and wheat. Many wheat breeding lines in the USA carry exotic genes for resistance to pathogens: powdery mildew (*Erysiphe graminis*); dwarf bunt (*Tilletia controversa*); several rust species (*Puccinia* spp); and leaf blotch (*Septoria nodorum*), and to insects: Hessian fly (*Mayetiola destructor*) and greenbug (*Schizaphis graminum*). Resistance to an important new pest, the Russian wheat aphid (*Diuraphis noxia*) has been found in about 50 accessions of wheat and barley (0.5% of the accessions screened) and most of these are landraces are from Afghanistan and the former Soviet Union.

But wild accessions and landraces that provide the most HPR genes are underrepresented in many germplasm collections. Only about 8% of accessions in the world crop germplasm collections are wild crop relatives and some collections have little or no wild germplasm. In most crops (with the exception of wheat, oats, potato, and tomato) it is estimated that less than 20% of the genetic diversity represented in wild species and unimproved cultivars is conserved *ex situ*. Some critics contend that even this estimate may be low because estimates of the total available variation are inaccurate. Time and resource pressures to obtain material has resulted in relative undersampling of geographically remote areas.

Although more germplasm collecting is needed and continues, preservation of the accessions may be problematic. Resources are already strained to maintain collections at their present size. Losses and erosion in extensive collections inevitably occur due to contamination, drift, errors in record keeping, physical facilities failures, inadvertent selection during increases, and other causes. Collections require backups at one or more locations to protect against these kinds of losses, but accessions in some crops do not have sufficient backups. In other crops, germplasm duplication is excessive among international collections, but coordinating optimal reductions is logistically difficult. Curators and managers of these collections are working hard to minimise these problems.

Another difficulty with extensive *ex situ* collections is that many accessions are inadequately characterised, or have undesirable traits that discourage breeders from using them. Evaluation and enhancement within the context of curation requires considerable resources. To address these problems with available resources many collections are now managed using the ëcoreí concept. A subset of accessions is chosen to maximise representation of the alleles in the entire collection. Unfortunately, a typical (10%) core will contain approximately 70% of the collection alleles and may miss rare alleles such as those conferring pest resistance. Meanwhile, devotion of resources to a core can lead to more rapid erosion of main collection along with its potential pest resistance genes. Available resources are simply not sufficient to preserve and characterise genetic variation *ex situ*.

IN SITU GENETIC CONSERVATION

One approach to preservation, if not characterisation, of crop genetic diversity is establishment of *in situ* reserves. These are preserved habitats harboring wild populations of crop species or related species; or they are regions where cultivation of indigenous landraces is continued by "custodian" farmers in exchange for subsides. *In situ* genetic reserves can help conserve wider genetic variation less expensively than *ex situ* conservation because the costs of germplasm storage are eliminated. Additionally, while *ex situ* collections are genetically static or may even erode, genetic variation of *in situ* populations should be maintained by natural processes or indigenous cultural practices. Accumulation of deleterious mutations will also be prevented in this way. Genetic and ecological patterns in reserve populations, e.g., pest resistance or microclimate tolerance, can be identified *in situ* and used to extract useful traits.

The advantages of *in situ* genetic conservation is widely recognised, but efforts in this area are just beginning. Geographic reserves have been established for a number of crops including apples, pears, pecan, palm species, forage crops, cassava, peanut, teosinte, and *Allium, Avena* and *Triticum* species. But in situ conservation involves more than just setting aside habitat reserves. To be useful, such reserves must be managed to document and maintain genetic variability, usually necessitating several ecogeographically distinct reserves for a particular species.

In situ reserves must be managed along with *ex situ* collections to facilitate extraction of useful genes. This level of in situ conservation does not exist for any crop or crop relative. The closest coordination between *ex situ* and in situ conservation may be the efforts of CIMMYT to oversee the establishment of teosinte geographic reserves. In fact, just how to accomplish effective in situ genetic conservation is still an area of study. The theory and methods are under development generating a complex scientific literature.

Agricultural Genetic Conservation and Global Biological Diversity

The application of molecular biology to crop breeding theoretically makes accessible the world gene pool. The spectacular success of insect resistance derived from genes of the soil bacterium *Bacillus thuringiensis* illustrates that any genome potentially can contribute to HPR and other invaluable crop traits. So agricultural genetic conservation and conservation of the global genetic diversity are becoming less distinct. Many of the theoretical considerations and methods are similar as well. In situ conservation approaches will become more important as global pressures reduce the size of natural habitat reserves to their theoretical minimum sizes necessary to maintain species viability. *Ex situ* reserves of some wild species will play a part in preservation and restoration of some habitats. Often, cooperative efforts between ecological and crop germplasm conservators will be appropriate. For example, some wildland habitats can serve as in situ genetic reserves for crop relatives.

The relationship between agriculture and global conservation is complex. Although agricultural productivity depends on global genetic diversity, and may provide some technology that can help preserve it, at the same time agriculture is one of the most environmentally destructive human activities. First, because of its scale ñ approximately 50% of the terrestrial surface is under cultivation. Second, because of its intensity, most agroecosystems largely supplant natural communities. Ironically, vascular plants, which carry the majority of genes with potential for crop improvement, are most severely affected by agricultural expansion. And agriculture is also harmful to genetic diversity because of pesticides effects on nontarget organisms.

Thus, although it is estimated that agricultural production may need to double to feed projected human populations during the next 25 years, increasing production can no longer be accomplished by increasing the area under cultivation. And the ancillary negative effects of agriculture on global ecology must be minimised. To accomplish these goals, genetic improvement of crop yields and tolerance to abiotic and biotic stresses (HPR) will be critical. But there is a potentially dangerous interdependence: as global biological diversity erodes, so does the genetic potential needed to conserve it.

Accomplishing Genetic Conservation

The imperatives for conserving needed genetic diversity outlined herein include expanding *ex situ* germplasm collections of wild crop relatives and unimproved varieties; improving accessibility to breeders of genes in unimproved accessions through better characterisation and enhancement; expanding use of in situ reserves; and reducing agricultureís negative effects on global biodiversity. Accomplishing these goals will require substantial resources; just adequately conserving the existing *ex situ* collections will require tripling of present funding. New techniques will help improve the cost-effectiveness of *ex situ* preservation. For example, cryogenic storage can increase the time germplasm can be stored without regeneration.

Newer molecular techniques for the characterisation of germplasm are faster and cheaper. In addition to the technological challenges, social, political and economic factors complicate genetic conservation. National and private proprietary issues can impede cooperation. Most diversity resides in undercapitalised countries that cannot afford to conserve it. Extraction and of this diversity by developed countries can easily become a form of economic exploitation. Unchecked market forces will result in unjust distribution of genetic resources and overexploitation. Strong international agreements are needed.

The International Convention on Biological Diversity (ICBD), signed by many heads of state after the Earth Summit in Rio de Janeiro acknowledged, at least in principle, the interelatedness of conservation and sustainable production, the need to channel international public and private resources towards conservation of biological diversity, the requirement to protect the interests of the less developed countries in which so much genetic diversity resides while providing incentives to attract investment in conservation from the more developed countries, and the requirement to somehow curb human use of global resources while assuring a good quality of life for everyone.

REFERENCES

Ehrenfeld, D., *The arrogance of humanism*, Oxford University Press, Oxford, 1981.

Ehrlich, P. R., Wilson, E. O., "Biodiversity studies: science and policy", *Science*, 1991.

Eigenbrode, S. D., J. T. Trumble & R. A. Jones, "Resistance to beet armyworm (Spodoptera exigua [Hubner]), hemipterans, and Liriomyza spp. in Lycopersicon", *Journal of the American Society of Horticultural Science*, 1993.

————————, "Plant resistance to insects in integrated pest management in vegetables", *Journal of Agricultural Entomology*, 1994.

8

Use of Insecticides

Insecticides are agents of chemical or biological origin that control insects. Control may result from killing the insect or otherwise preventing it from engaging in behaviors deemed destructive. Insecticides may be natural or manmade and are applied to target pests in a myriad of formulations and delivery systems (sprays, baits, slow-release diffusion, etc.). The science of biotechnology has, in recent years, even incorporated bacterial genes coding for insecticidal proteins into various crop plants that deal death to unsuspecting pests that feed on them.

Some 10,000 species of the more than 1 million species of insects are crop-eating, and of these, approximately 700 species worldwide cause most of the insect damage to man's crops, in the field and in storage. Humanoids have been on earth for more than 3 million years, while insects have existed for at least 250 million years. We can guess that among the first approaches used by our primitive ancestors to reduce insect annoyance was hugging smoky fires or spreading mud and dust over their skin to repel biting and tickling insects, a practice resembling the habits of elephants, swine, and water buffalo. Today, such approaches would be classed as *repellents,* a category of *insecticides.*

Historians have traced the use of pesticides to the time of Homer around 1000 B.C., but the earliest records of insecticides pertain to the burning of "brimstone" (sulfur) as a fumigant. Pliny the Elder recorded most of the earlier insecticide uses in his *Natural History*. Included among these were the use of gall from a green lizard to protect apples from worms and rot. Later, we find a variety of materials used with questionable results: extracts of pepper and tobacco, soapy water, whitewash, vinegar, turpentine, fish oil, brine, lye among many others.

At the beginning of World War II, our insecticide selection was limited to several arsenicals, petroleum oils, nicotine, pyrethrum, rotenone, sulfur, hydrogen cyanide gas, and cryolite. It was World War II that opened the *Modern Era of Chemical* control with the introduction of a new concept of insect control —synthetic organic insecticides, the first of which was DDT.

ORGANOCHLORINES

The organochlorines are insecticides that contain carbon (thus *organo-*), hydrogen, and chlorine. They are also known by other names: *chlorinated hydrocarbons, chlorinated organics, chlorinated insecticides,* and *chlorinated synthetics.* The organochlorines are now primarily of historic interest, since few survive in today's arsenal.

Diphenyl Aliphatics

The oldest group of the organochlorines is the *diphenyl aliphatics,* which included DDT, DDD, dicofol, ethylan, chlorobenzilate, and methoxychlor. DDT is probably the best known and most notorious chemical of the 20th century. It is also fascinating, and remains to be acknowledged as the most useful insecticide developed. More than 4 billion pounds of DDT were used throughout the world, beginning in 1940, and in the U.S. ending essentially in 1973, when the U.S. Environmental Protection Agency canceled all uses. The remaining First World countries rapidly followed suit. DDT is still effectively used for malaria control in several third world countries. In 1948, Dr. Paul Muller, a Swiss entomologist, was awarded the Nobel Prize in Medicine for his lifesaving discovery of DDT as an insecticide useful in the control of malaria, yellow fever and many other insect-vectored diseases.

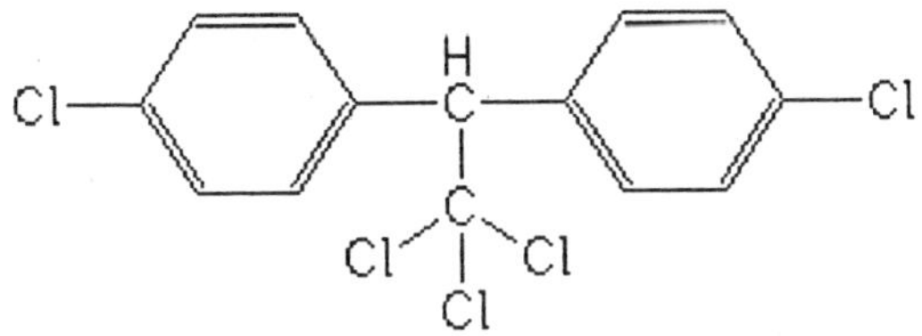

Mode of action

The mode of action for DDT has never been clearly established, but in some complex manner it destroys the delicate balance of sodium and potassium ions within the axons of the neuron in a way that prevents normal transmission of nerve impulses, both in insects and mammals. It apparently acts on the sodium channel to cause "leakage" of sodium ions. Eventually the affected neurons fire impulses spontaneously, causing the muscles to twitch— "DDT jitters"— followed by convulsions and death. DDT has a negative temperature correlation—the lower the surrounding temperature the more toxic it becomes to insects.

Hexchlorocyclohexane (HCH)

Also known as benzenehexachloride (BHC), the insecticidal properties of HCH were discovered in 1940 by French and British entomologists. In its technical grade, there are five isomers, *alpha, beta, gamma, delta* and *epsilon*. Surprisingly, only the *gamma* isomer has insecticidal properties. Consequently, the *gamma* isomer was isolated in manufacture and sold as the odorless insecticide *lindane*. In contrast, technical grade HCH has a strong musty odor and flavor, which can be imparted to treated crops and animal products. Because of its very low cost, HCH is still used in many developing countries. In 2002, the U.S. EPA removed all food-related (tolerance-requiring) uses of lindane from the U.S.

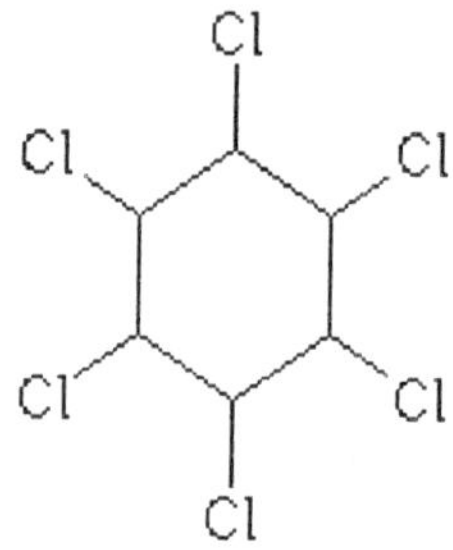

Mode of action

The effects of HCH superficially resemble those of DDT, but occur much more rapidly, and result in a much higher rate of respiration in insects. The *gamma* isomer is a neurotoxicant whose effects are normally seen within hours as increased activity, tremors, and convulsions leading to prostration. It too, exhibits a negative temperature correlation, but not as pronounced as that of DDT.

Cyclodienes

The cyclodienes appeared after World War II: chlordane, 1945, aldrin and dieldrin, 1948; heptachlor, 1949; endrin, 1951; mirex, 1954; endosulfan, 1956; and chlordecone, 1958. There were other cyclodienes of minor importance developed in the U.S. and Germany. Most of the cyclodienes are persistent insecticides and are stable in soil and relatively stable to the ultraviolet of sunlight. As a result, they were used in greatest quantity as soil insecticides (especially chlordane, heptachlor, aldrin, and dieldrin) for the control of termites and soil-borne insects whose larval stages feed on the roots of plants.

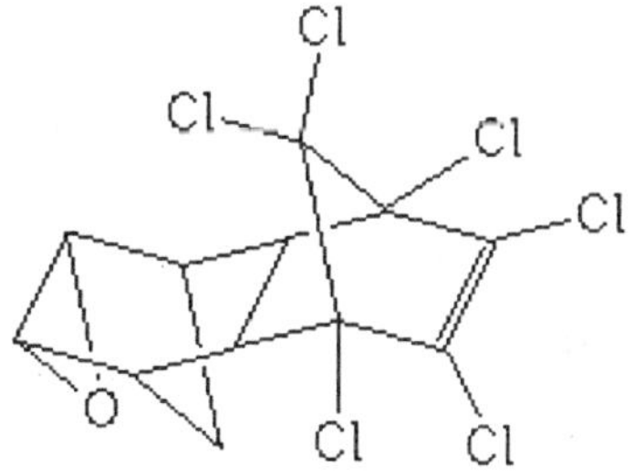

To appreciate the effectiveness of these materials as termiticides, consider that wood and wooden structures treated with chlordane, aldrin, and dieldrin in the year of their development are still protected from damage—after more than 60 years! The cyclodienes were the most effective, long-lasting and economical termiticides ever developed. Because of their persistence in the environment, resistance that developed in several soil insect pests, and in some instances *biomagnification* in wildlife food chains, most agricultural uses of cyclodienes were canceled by the EPA between 1975 and 1980, and their use as termiticides canceled in 1984-88.

Mode of action

Unlike DDT and HCH, the cyclodienes have a positive temperature correlation—their toxicity increases with increasing ambient temperature. Their modes of action are also not clearly understood. However, it is known that this group acts on the inhibitory mechanism called the GABA (g-aminobutyric acid) receptor. This receptor operates by increasing chloride ion permeability of neurons. Cyclodienes prevent chloride ions from entering the neurons, and thereby antagonise the "calming" effects of GABA. Cyclodienes appear to affect all animals similarly, first with the nervous activity followed by tremors, convulsions and prostration.

Polychloroterpenes

Only two polychloroterpenes were developed—toxaphene in 1947, and strobane in 1951. Toxaphene had by far the greatest use of any single insecticide in agriculture, while strobane was relatively insignificant. Toxaphene was used on cotton, first in combination with DDT, for alone it had minimal insecticidal qualities. Then, in 1965, after several major cotton insects became resistant to DDT, toxaphene was formulated with methyl parathion, an organophosphate insecticide mentioned later.

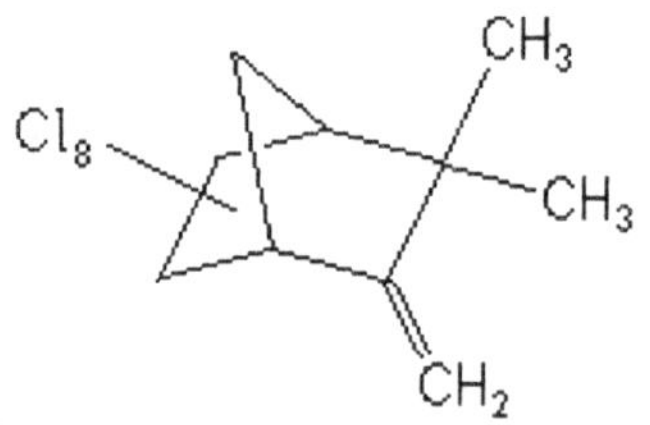

Toxaphene is a mixture of more than 177 10-carbon polychlorinated derivatives. These materials persist in the soil, though not as long as the cyclodienes, and disappear from the surfaces of plants in 3-4 weeks. This disappearance is attributed more to volatility than to photolysis or plant metabolism. Toxaphene is rather easily metabolised by mammals and birds, and is not stored in body fat nearly to the extent of DDT, HCH and the cyclodienes. Despite its low toxicity to insects, mammals and birds, fish are highly susceptible to toxaphene poisoning, in the same order of magnitude as to the cyclodienes. Toxaphene's registrations were canceled by EPA in 1983.

Mode of action

Toxaphene and strobane act on the neurons, causing an imbalance in sodium and potassium ions, similar to that of the cyclodiene insecticides.

Organophosphates

Organophosphates (OPs) is the term that includes all insecticides containing phosphorus. Other names used, but no longer in vogue, are *organic phosphates, phosphorus insecticides, nerve gas relatives,* and *phosphoric acid esters.* All organophosphates are derived from one of the phosphorus acids, and as a class are generally the most toxic of all pesticides to vertebrates. Because of the similarity of OP chemical structures to the "nerve gases," their modes of action are also similar. Their insecticidal qualities were first observed in Germany during World War II in the study of the extremely toxic OP nerve gases *sarin, soman,* and *tabun.* Initially, the discovery was made in search of substitutes for nicotine, which was heavily used as an insecticide but in short supply in Germany.

The OPs have two distinctive features: they are generally more toxic to vertebrates than other classes of insecticides, and most are chemically unstable or nonpersistent. It is this latter characteristic that brought them into agricultural use as substitutes for the persistent *organochorines.* Because of the relatively high toxicity of the OP's, EPA, under provisions of the Food Quality Protection Act, undertook an extensive reappraisal of the entire class beginning in the late 1990's. Many OP's were voluntarily canceled and others lost uses.

Mode of action

The OPs work by inhibiting certain important enzymes of the nervous system, namely *cholinesterase* (ChE). The enzyme is said to be *phosphorylated* when it becomes attached to the phosphorous moiety of the insecticide, a binding that is irreversible. This inhibition results in the accumulation of acetylcholine (ACh) at the neuron/neuron and neuron/muscle (neuromuscular) junctions or synapses, causing rapid twitching of voluntary muscles and finally paralysis.

Classification

All OPs are esters of phosphorus having varying combinations of oxygen, carbon, sulfur and nitrogen attached, resulting in six different subclasses: phosphates, phospho-nates, phosphorothioates, phosphorodithioates, phosphorothiolates and phosphoramidates. These subclasses are easily identified by their chemical names.

The OPs are generally divided into three groups—*aliphatic, phenyl,* and *heterocyclic* derivatives.

Aliphatics

The aliphatic OPs are carbon chain-like in structure. The first OP brought to agriculture, TEPP (1946) belonged to this group. Other examples are malathion, trichlorfon (Dylox®), monocrotophos (Azodrin®), dimethoate (Cygon®), oxydemetonmethyl (Meta Systox®), dicrotophos (Bidrin®), disulfoton (Di-Syston®), dichlorvos (Vapona®), mevinphos (Phosdrin®), methamidophos (Monitor®), and acephate (Orthene®).

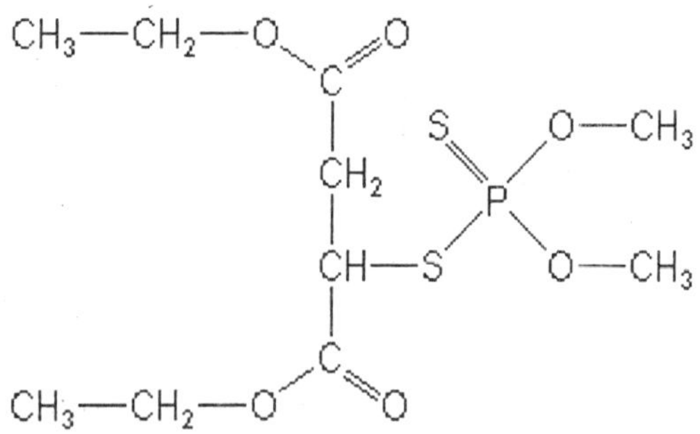

Phenyl Derivatives

The phenyl OPs contain a phenyl ring with one of the ring hydrogens displaced by attachment to the phosphorus moiety and other hydrogens frequently displaced by Cl,

NO_2, CH_3, CN, or S. The phenyl OPs are generally more stable than the aliphatics, thus their residues are longer lasting. The first phenyl OP brought into agriculture was parathion (ethyl parathion) in 1947. Examples of other phenyl OPs are methyl parathion, profenofos (Curacron®), sulprofos (Bolstar®), isofenphos (Oftanol®, Pryfon®), fenitrothion (Sumithion®), fenthion (Dasanit®), and famphur (Cyflee®. Warbex®).

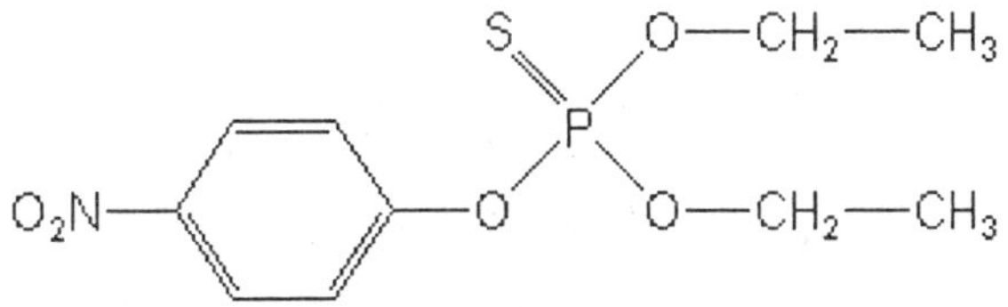

Heterocyclic Derivatives

The term *heterocyclic* means that the ring structures are composed of different or unlike atoms, e.g.,oxygen, nitrogen or sulfur. The first of this group was diazinon introduced in 1952. Other examples in this group are azinphos-methyl (Guthion®), azinphos-ethyl (Acifon®, Gusathion®), chlorpyrifos (Dursban®, Lorsban®, Lock-On®), methidathion (Supracide®), phosmet (Imidan®), isazophos (Brace®, Triumph®), and chlorpyrifos-methyl (Reldan®).

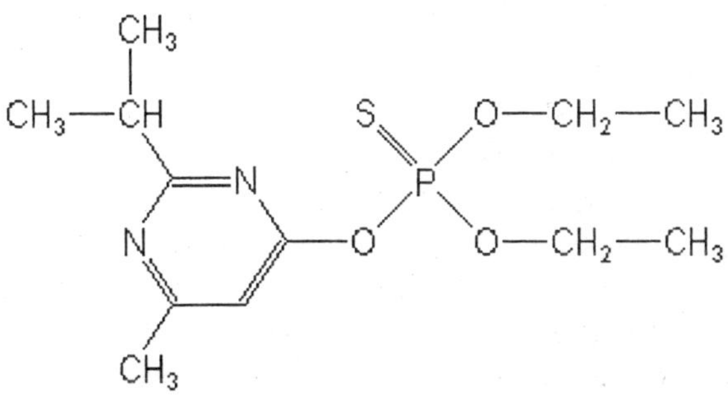

ORGANOSULFURS

These few materials have very low toxicity to insects and are used only as acaricides (miticides). They contain two phenyl rings, resembling DDT, with sulfur in place of carbon as the central atom. These include tetradifon (Tedion®), propargite (Omite®, Comite®), and ovex (Ovotran®).

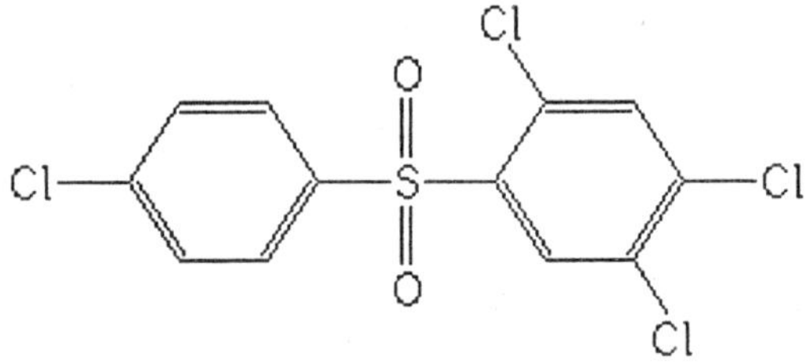

CARBAMATES

The carbamate insecticides are derivatives of carbamic acid (as the OPs are derivatives of phosphoric acid). And like the OPs, their mode of action is that of inhibiting the vital enzyme *cholinesterase* (ChE).

The first successful carbamate insecticide, carbaryl (Sevin®), was introduced in 1956. More of it has been used worldwide than all the remaining carbamates combined. Two distinct qualities have made it the most popular carbamate: its very low mammalian oral and dermal toxicity and an exceptionally broad spectrum of insect control. Other long-standing carbamate insecticides are methomyl (Lannate®), carbofuran (Furadan®), aldicarb (Temik®), oxamyl (Vydate®), thiodicarb (Larvin®), methiocarb (Mesurol®), propoxur (Baygon®), bendiocarb (Ficam®), carbosulfan (Advantage®), aldoxycarb (Standak®), promecarb (Carbamult®), and fenoxycarb (Logic®, Torus®). Carbamates more recently introduced include primicarb, indoxacarb, alanycarb and furathiocarb.

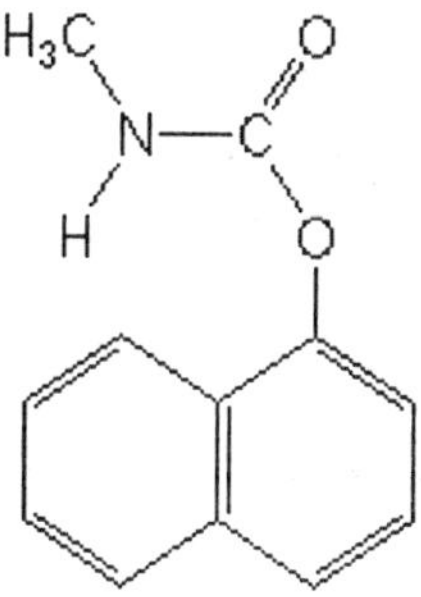

Mode of Action

Carbamates inhibit cholinesterase (ChE) as OPs do, and they behave in almost identical manner in biological systems, but with two main differences. Some carbamates are potent inhibitors of aliesterase (miscellaneous aliphatic esterases whose exact functions are not known), and their selectivity is sometimes more pronounced against the ChE of different species. Second, ChE inhibition by carbamates is reversible. When ChE is inhibited by a carbamate, it is said to be *carbamylated*, as when an OP results in the enzyme being *phosphorylated*. In insects, the effects of OPs and carbamates are primarily those of poisoning of the central nervous system, since the insect neuromuscular junction is not cholinergic, as in mammals. The only cholinergic synapses known in insects are in the central nervous system. (The chemical neuromuscular junction transmitter in insects is thought to be glutamic acid.)

FORMAMIDINES

The formamidines comprise a small group of insecticides. Three examples are chlordimeform (Galecron®, Fundal®), which is no longer registered in the U.S., formetanate (Carzol®), and amitraz (Mitac®, Ovasyn®). Their current value lies in the control of OP- and carbamate-resistant pests.

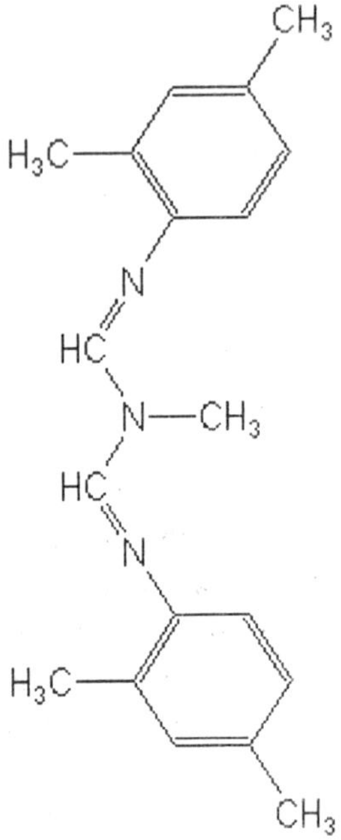

Mode of Action

Formamidine poisoning symptoms are distinctly different from other insecticides. Their proposed action is the inhibition of the enzyme monoamine oxidase, which is responsible for degrading the neurotransmitters norepinephrine and serotonin. This results in the accumulation of these compounds, which are known as *biogenic amines*. Affected insects become quiescent and die.

DINITROPHENOLS

The basic dinitrophenol molecule has a broad range of toxicities—as herbicides, insecticides, ovicides, and fungicides. Of the insecticides, binapacryl (Morocide®) and dinocap (Karathane®) were the most recently used. Dinocap is an effective miticide and was very heavily used as a fungicide for the control of powdery mildew fungi.

Because of the inherent toxicity of the dinitrophenols, they have all been withdrawn.

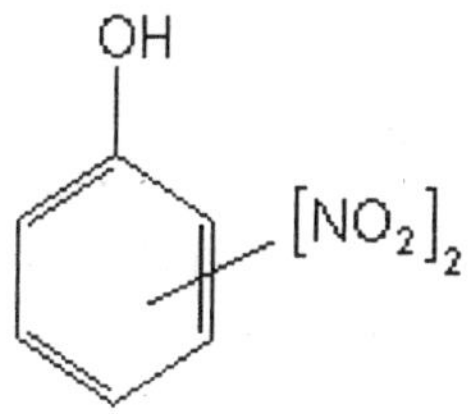

Mode of Action

Dinitrophenols act by uncoupling or inhibiting oxidative phosphorylation, which basically prevents the formation of the high-energy phosphate molecule, adenosine triphosphate (ATP).

ORGANOTINS

The organotins are a group of acaricides that double as fungicides. Of particular interest is cyhexatin (Plictran®), one of the most selective acaricides known, introduced in 1967. Fenbutatin-oxide (Vendex®) has been used extensively against mites on deciduous fruits, citrus, greenhouse crops, and ornamentals.

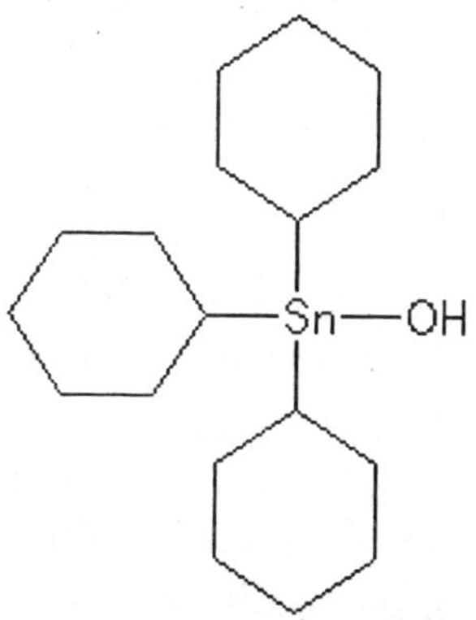

Mode of Action

These tin compounds inhibit oxidative phosphorylation at the site of dinitrophenol uncoupling, preventing the formation of the high-energy phosphate molecule adenosine triphosphate (ATP). These trialkyl tins also inhibit photophosphorylation in chloroplasts, the chlorophyll-bearing subcellular units) and could therefore serve as algicides.

PYRETHROIDS

Natural pyrethrum has seldom been used for agricultural purposes because of its cost and instability in sunlight. In recent decades, many synthetic pyrethrin-like materials have become available. They were originally referred to as *synthetic pyrethroids*. Currently, the better nomenclature is simply *pyrethroids*. These are stable in sunlight and are generally effective against most agricultural insect pests when used at the very low rates of 0.01 to 0.1 pound per acre.

The pyrethroids have an interesting evolution, which is conveniently divided into four generations. The first generation contains only one pyrethroid, allethrin (Pynamin®), which appeared in 1949. Its synthesis was very complex, involving 22 chemical reactions to reach the final product.

The second generation includes tetramethrin (Neo-Pynamin®) (1965), followed by resmethrin (Synthrin®) in 1967 (20X as effective as pyrethrum), then bioresmethrin (50X as effective as pyrethrum) (1967), then Bioallethrin® (1969), and finally phonothrin (Sumithrin®) (1973).

The third generation includes fenvalerate (Pydrin®, Tribute®, & Bellmark®), and permethrin (Ambush®, Astro®, Dragnet®, Flee®, Pounce®, Prelude®, Talcord®& Torpedo®) which appeared in 1972-73. These became the first agricultural pyrethroids because of their exceptional insecticidal activity (0.1 lb ai/A) and their photostability. They were virtually unaffected by ultraviolet in sunlight, lasting 4-7 days as efficacious residues on crop foliage.

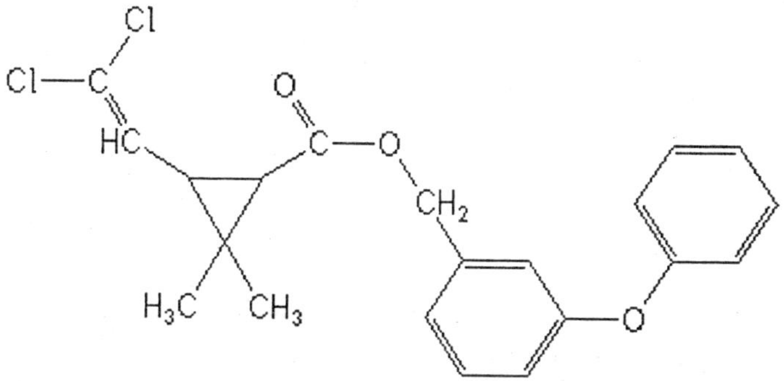

The fourth and current generation, is truly exciting because of their effectiveness in the range of 0.01 to 0.05 lb ai/A. These include bifenthrin (Capture®, Talstar®), *lambda*-cyhalothrin (Demand®, Karate®, Scimitar®& Warrior®), cypermethrin (Ammo®, Barricade®, Cymbush®, Cynoff® & Ripcord®), cyfluthrin (Baythroid®, Countdown®, Cylense®, Laser® & Tempo®), deltamethrin (Decis®) esfenvalerate (Asana®, Hallmark®), fenpropathrin (Danitol®), flucythrinate (Cybolt®, Payoff®), fluvalinate (Mavrik®; Spur ®, discontinued), prallethrin (Etoc®), *tau*-fluvalinate (Mavrik®) tefluthrin (Evict®, Fireban®, Force® & Raze®), tralomethrin (Scout X-TRA®, Tralex®), and *zeta*-cypermethrin (Mustang® & Fury®). All of these are photostable, that is, they do not undergo photolysis (splitting) in sunlight. And because they have minimal volatility they provide extended residual effectiveness, up to 10 days under optimum conditions.

Recent additions to the fourth generation pyrethroids are acrinathrin (Rufast®), imiprothrin (Pralle®), registered in 1998, and *gamma*-cyhalothrim (Pytech®), which is in development.

Mode of Action

The pyrethroids share similar modes of action, resembling that of DDT, and are considered axonic poisons. They apparently work by keeping open the sodium channels in neuronal membranes. There are two types of pyrethroids. Type I, among other physiological responses, have a negative temperature coefficient, resembling that of DDT. Type II, in contrast have a positive temperature coefficient, showing increased kill with increase in ambient temperature. Pyrethroids affect both the peripheral and central nervous system of the insect. They initially stimulate nerve cells to produce repetitive discharges and eventually cause paralysis. Such effects are caused by their action on the sodium channel, a tiny hole through which sodium ions are permitted to enter the axon to cause excitation. The stimulating effect of pyrethroids is much more pronounced than that of DDT.

NICOTINOIDS

The nicotinoids are a newer class of insecticides with a new mode of action. They have been previously referred to as *nitro-quanidines, neonicotinyls, neonicotinoids, chloronicotines,* and more recently as the *chloronicotinyls*. Just as the synthetic pyrethroids are similar to and modeled after the natural pyrethrins, so too, are the nicotinoids similar to and modeled after the natural nicotine–. Imidacloprid was introduced in Europe and Japan in 1990 and first registered in the U.S. in 1992. It is currently marketed as several proprietary products worldwide, e.g., Admire®, Confidor®, Gaucho®, Merit®, Premier®, Premise® and Provado®. Very possibly it is used in the greatest volume globally of all insecticides.

Imidacloprid is a systemic insecticide, having good root-systemic characteristics and notable contact and stomach action. It is used as a soil, seed or foliar treatment in cotton, rice cereals, peanuts, potatoes, vegetables, pome fruits, pecans and turf, for the control of sucking insects, soil insects, whiteflies, termites, turf insects and the Colorado potato beetle, with long residual control. Imidacloprid has no effect on mites or nematodes.

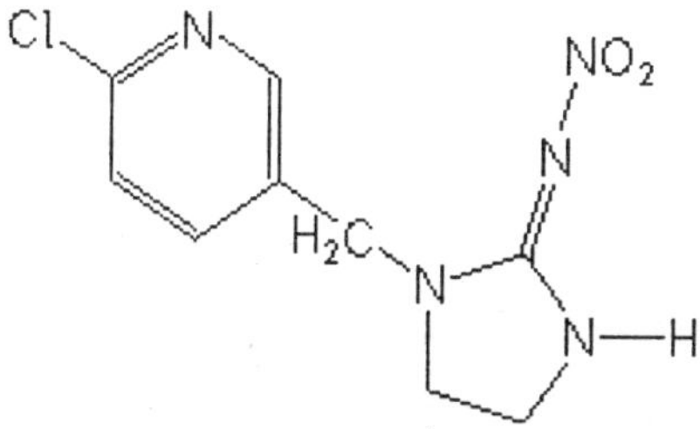

Other nicotinoids include acetamiprid (Assail®), thiamethoxam (Actara®, Platinum®), nitenpyram (Bestguard®), clothianidin (Poncho®, dinotefuran (Starke®) and thiacloprid. U.S. registrations for acetamiprid, thiamethoxam and thiacloprid were granted in 2002 and clothianidin in 2003.

Mode of Action

The nicotinoids act on the central nervous system of insects, causing irreversible blockage of postsynaptic nicotinergic acetylcholine receptors.

SPINOSYNS

Spinosyns are among the newest classes of insecticides, represented by spinosad (Success®, Tracer Naturalyte®). Spinosad is a fermentation metabolite of the actinomycete *Saccharopolyspora spinosa*, a soil-inhabiting microorganism. It has a novel molecular structure and mode of action that provide excellent crop protection typically associated with synthetic insecticides, first registered for use on cotton in 1997. Spinosad is a mixture of spinosyns A and D (thus its name, spinosAD). It is particularly effective as a broad-spectrum material for most caterpillar pests at the astonishing rates of 0.04 to 0.09 pound of active ingredient (18 to 40 grams) per acre. It has both contact and stomach activity against lepidopteran larvae, leaf miners, thrips, and termites, with long residual activity. Crops registered include cotton, vegetables, tree fruits, ornamentals and others.

Mode of Action

Spinosad acts by disrupting binding of acetylcholine in nicotinic acetylcholine receptors at the postsynaptic cell.

FIPROLES (OR PHENYLPYRAZOLES)

Fipronil (Regent®, Icon®, Frontline®) is the only insecticide in this new class, introduced in 1990 and registered in the U.S. in 1996. It is a systemic material with contact and stomach activity.

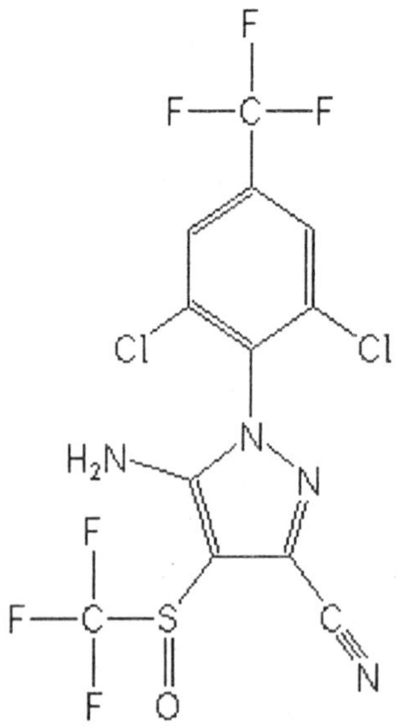

Fipronil is used for the control of many soil and foliar insects, (e.g., corn rootworm, Colorado potato beetle, and rice water weevil) on a variety of crops, primarily corn, turf, and for public health insect control. It is also used for seed treatment and formulated as baits for cockroaches, ants and termites. Fipronil is effective against insects resistant or tolerant to pyrethroid, organophosphate and carbamate insecticides.

Mode of Action

Fipronil blocks the (g-aminobutyric acid- (GABA) regulated chloride channel in neurons, thus antagonising the "calming" effects of GABA, similar to the action of the Cyclodienes.

PYRROLES

Chlorfenapyr (Alert®, Pirate®) is the first and only member of this unique chemical group, as both a contact and stomach insecticide-miticide. It is used on cotton and experimentally on corn, soybeans, vegetables, tree and vine crops, and ornamentals to control whitefly, thrips, caterpillars, mites, leafminers, aphids, and Colorado potato beetle. It has ovicidal activity on some species. EPA took the unusual step of refusing to register chlorfenapyr in 2000 for cotton insect control because of potential hazards to birds. However, labels for greenhouse ornamentals were granted in 2001.

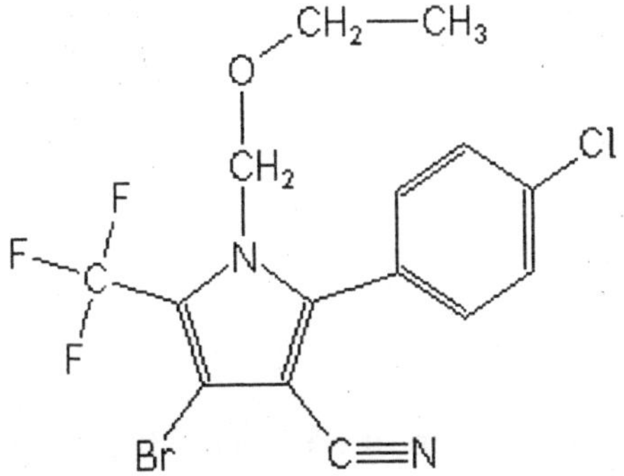

Mode of Action

Chlorfenapyr is an "uncoupler" or inhibitor of oxidative phosphorylation, preventing the formation of the crucial energy molecule adenosine triphosphate (ATP).

PYRAZOLES

The original pyrazoles were tebufenpyrad and fenpyroximate (not illustrated). These were designed primarily as non-systemic contact and stomach miticides, but do have limited effectiveness on psylla, aphids, whitefly, and thrips. Tebufenpyrad (Pyranica®, Masai®), registered by EPA in 2002, is used on cotton, soybeans, vegetables, pome fruits, grapes and citrus. Fenpyroximate (Acaban®, Dynamite®) controls all stages of mites, gives fast knockdown, inhibits molting of immature stages of mites, and has long residual activity. Newer members of this class include ethiprole (Curbex®) which is active on a broad sprectum of chewing and sucking insects, and tolfenpyrad (OMI-88) which is reputed to active on pests infesting cole and cucurbit crops.

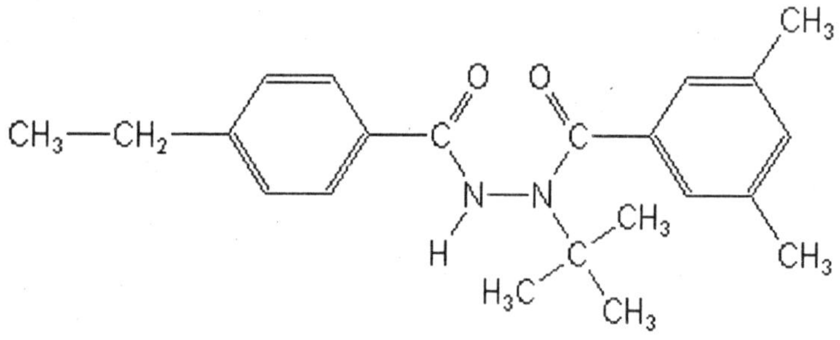

Mode of Action

Their mode of action is that of inhibiting mitochondrial electron transport at the NADH-CoQ reductase site, leading to the disruption of adenosine triphosphate (ATP) formation, the crucial energy molecule.

PYRIDAZINONES

Pyridaben (Nexter®, Sanmite®) is the only member of this class. It is a selective contact insecticide and miticide, also effective against thrips, aphids, whiteflies and leafhopprs. Registrations are for pome fruits, almonds, citrus, ornamentals and greenhouse ornamentals. Pyridaben provides exceptionally long residual control, and rapid knockdown at a broad range of temperatures.

Mode of action—Pyridaben is a metabolic inhibitor that interrupts mitochondrial electron transport at Site 1, similar to the Quinazolines, below.

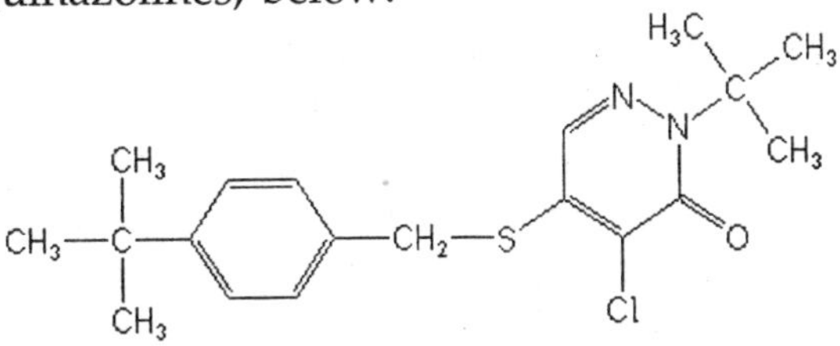

QUINAZOLINES

The quinazolines offer a unique chemical configuration, consisting only of one insecticide, fenazaquin (Matador®). Fenazaquin is a contact and stomach miticide. It has ovicidal activity, gives rapid knockdown, and controls all stages of mites. Not yet registered in the U.S., it is used on cotton, stone and pome fruits, citrus, grapes and ornamentals.

FENAZAQUIN (MATADOR®)

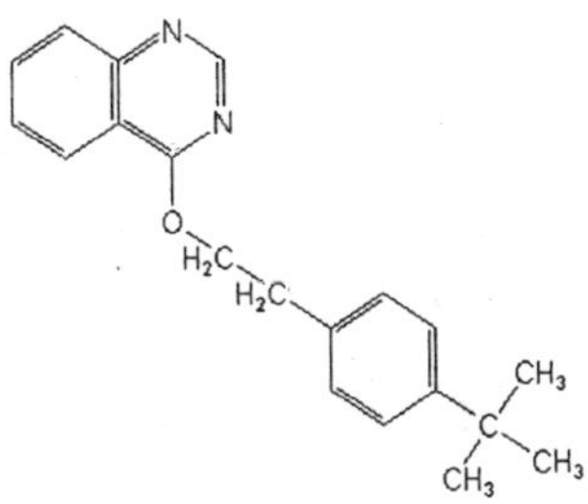

Mode of Action

Fenazaquin inhibits mitochondrial electron transport at Site 1, similar to the Pyridazinones, above.

BENZOYLUREAS

Benzoylureas are an entirely different class of insecticides that act as insect growth regulators (IGRs). Rather than being the typical poisons that attack the insect nervous system, they interfere with chitin synthesis and are taken up more by ingestion than by contact. Their greatest value is in the control of caterpillars and beetle larvae.

Benzoylureas were first used in Central America in 1985, to control a severe, resistant leafworm complex (*Spodoptera spp., Trichoplusia spp.*) outbreak in cotton. The withdrawal of the ovicide chlordimeform made their control quite difficult due to their high resistance to almost all insecticide classes, including the pyrethroids.

The benzoylureas were introduced in 1978 by Bayer of Germany, triflumuron (Alsystin®) being the first. Others appearing since then are chlorfluazuron (Atabron®), followed by teflubenzuron (Nomolt®, Dart®), hexaflumuron (Trueno®, Consult®), flufenoxuron (Cascade®), and flucycloxuron (Andalin®). Others are flurazuron, novaluron, and diafenthiuron, bistrifluron (DBI-3204) and noviflumuron (XDE-007). Until recently lufenuron (Axor®) was the newest addition to this group, appearing in 1990. Among the newer benzoylureas only hexaflumuron and novaluron have been registered by EPA.

The only other benzoylurea registered in the U.S. is diflubenzuron (Dimilin®, Adept®, Micromite®). It was first registered in 1982 for gypsy moth, cotton boll weevil, most forest caterpillars, soybean caterpillars, and mushroom flies, but now with a much broader range of registrations.

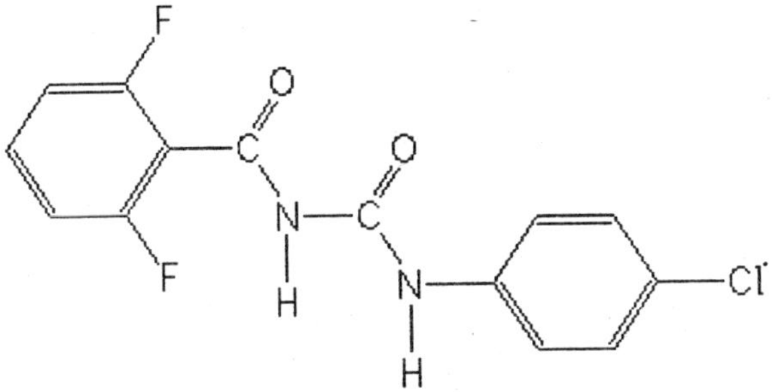

Though not a benzoylurea, cyromazine (Larvadex®, Trigard®), a triazine, is also a potent chitin synthesis inhibitor. It is selective toward Dipterous species and used for the control of leafminers in vegetable crops and ornamentals, and fed to poultry or sprayed to control flies in manure of broiler and egg producing operations, and incorporated into compost of mushroom houses for fungus gnats.

Mode of Action

The benzoylureas act on the larval stages of most insects by inhibiting or blocking the synthesis of chitin, a vital and almost indestructible part of the insect exoskeleton. Typical effects on developing larvae are the rupture of malformed cuticle or death by starvation. Adult female boll weevils exposed to diflubenzuron lay eggs that do not hatch. And, mosquito larvae control can be achieved with as little as 1.0 gram of diflubenzuron per acre of surface water.

BOTANICALS

Botanical insecticides are of great interest to many, for they are *natural* insecticides, toxicants derived from plants. Historically, the plant materials have been in use longer than any other group, with the possible exception of sulfur. Tobacco, pyrethrum, derris, hellebore, quassia, camphor, and turpentine were some of the more important plant products in use before the organized search for insecticides began in the early 1940s.

In recent years the term *biorational* has been put into play by the EPA. There are similarities and differences between the terms botanical and biorational. We include a section on the biorationals below and will address the overlaps in that section.

Botanical insecticide use in the U.S. peaked in 1966, and has declined steadily since. Pyrethrum is now the only classical botanical of significance in use. Some newer plant-derived insecticides that have come into use are referred to as *florals or scented plant chemicals* and include, among others, limonene, cinnamaldehyde and eugenol. In addition, there is azadirachtin from the neem tree which is used in greenhouse and on ornamentals.

PYRETHRUM

Pyrethrum is extracted from the flowers of a chrysanthemum grown in Kenya and Ecuador. It is one of the oldest and safest insecticides available. The ground, dried flowers were used in the early 19th century as the original louse powder to control body lice in the Napoleonic Wars. Pyrethrum acts on insects with phenomenal speed causing immediate paralysis, thus its popularity in fast knockdown household aerosols. However, unless it is formulated with one of the *synergists*, most of the paralysed insects recover to once again become pests. Pyrethrum is a mixture of four compounds: pyrethrins I and II and cinerins I and II.

Mode of Action

Pyrethrum is an axonic poison, as are the synthetic pyrethroids and DDT. Axonic poisons are those that in some way affect the electrical impulse transmission along the axons, the elongated extensions of the neuron cell body.

Pyrethrum and some pyrethroids have a greater insecticidal effect when the temperature is lowered, a negative temperature coefficient, as does DDT. They affect both the peripheral and central nervous system of the insect. Pyrethrum initially stimulates nerve cells to produce repetitive discharges, leading eventually to paralysis. Such effects are caused by their action on the sodium channel, a tiny hole through which sodium ions are permitted to enter the axon to cause excitation. These effects are produced in insect nerve cord, which contains ganglia and synapses, as well as in giant nerve fiber axons.

NICOTINE

Nicotine is extracted by several methods from tobacco, and is effective against most all types of insect pests, but is used particularly for aphids and caterpillars—soft bodied insects. Nicotine is an alkaloid, a chemical class of heterocyclic compounds containing nitrogen and having prominent physiological properties. Other well-known alkaloids that are not insecticides are caffeine (coffee, tea), quinine (cinchona bark), morphine (opium poppy), cocaine (coca leaves), ricinine (a poison in castor oil beans), strychnine (*Strychnos nux vomica*), coniine (spotted hemlock, the poison used by Socrates), and, finally LSD (a hallucinogen from the ergot fungus attacking grain).

Mode of Action

Nicotine action is one of the first, classic modes of action identified by pharmacologists. Drugs that act similarly to nicotine are said to have a nicotinic response. Nicotine mimics acetylcholine (ACh) at the neuromuscular (nerve/muscle) junction in mammals, and results in twitching, convulsions, and death, all in rapid order. In insects the same action is observed, but only in the central nervous system ganglia.

ROTENONE

Rotenone or rotenoids are produced in the roots of two genera of the legume family: *Derris* and *Lonchocarpus* (also called cubé) grown in South America. It is both a stomach and contact insecticide and used for the last century and a half to control leaf-eating caterpillars, and three centuries prior to that in South America to paralyse fish, causing them to surface and be easily captured. Today, rotenone is used in the same way to reclaim lakes for game fishing. Used on a prescribed basis, it eliminates all fish, closing the lake to reintroduction of rough species. It is a selective piscicide in that it kills all fish at dosages that are relatively nontoxic to fish food organisms, and is degraded rapidly.

Mode of Action

Rotenone is a respiratory enzyme inhibitor, acting between NAD+ (a coenzyme involved in oxidation and reduction in metabolic pathways) and coenzyme Q (a respiratory enzyme responsible for carrying electrons in some electron transport chains), resulting in failure of the respiratory functions.

LIMONENE

Limonene or *d*-Limonene is the latest addition to the botanicals. Limonene belongs to a group often called *florals* or *scented plant chemicals*. Extracted from citrus peel, it is effective against all external pests of pets, including fleas, lice, mites, and ticks, and is virtually nontoxic to warm-blooded animals. Several insecticidal substances occur in citrus oil, but the most important is limonene, which constitutes about 98% of the orange peel oil by weight. Two other recently introduced floral products are eugenol (Oil of Cloves) and cinnamaldehyde (derived from Ceylon and Chinese cinnamon oils). They are used on ornamentals and many crops to control various insects.

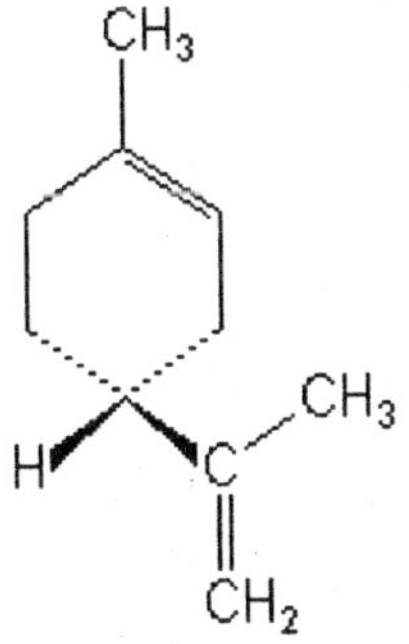

Mode of Action

Its mode of action is similar to that of pyrethrum. It affects the sensory nerves of the peripheral nervous system, but it is not a ChE inhibitor.

NEEM OIL

Neem oil extracts are squeezed from the seeds of the neem tree and contain the active ingredient *azadirachtin*, a nortriterpenoid belonging to the lemonoids. Azadirachtin has shown some rather sensational insecticidal, fungicidal and bactericidal properties, including insect growth regulating qualities. Azatin® is marketed as an insect growth regulator, and Align® and Nemix® as a stomach/ contact insecticide for greenhouse and ornamentals.

Mode of Action

Azadirachtin disrupts molting by inhibiting biosynthesis or metabolism of ecdysone, the juvenile molting hormone.

ANTIBIOTICS

In this category belong the *avermectins*, which are insecticidal, acaricidal, and antihelminthic agents that have been isolated from the fermentation products of

Streptomyces avermitilis, a member of the actinomycete family. *Abamectin* is the common name assigned to the avermectins, a mixture of containing 80% avermectin B1*a* and 20% B1*b*, homologs that have about equal biological activity. Clinch®-is a fire ant bait, and Avid®-is applied as a miticide/insecticide. Abamectin has certain local systemic qualities, permitting it to kill mites on a leaf's underside when only the upper surface is treated. The most promising uses for these materials are the control of spider mites, leafminers and other difficult-to-control greenhouse pests, and internal parasites of domestic animals.

EMAMECTIN BENZOATE

Emamectin benzoate (Proclaim®, Denim®) is an analog of abamectin, produced by the same fermentation system as abamectin. It was first registered in 1999. It is both a stomach and contact insecticide used primarily for control of caterpillars at the rate of 0.0075 to 0.015 lb (3.5 to 7.0 grams) a.i. per acre. Shortly after exposure, larvae stop feeding and become irreversibly paralysed, dying in 3-4 days. Rapid photodegradation of both abamectin and emamectin occurs on the leaf surface. More recently, milabectin (Mesa®) has been introduced. It is a miticide with activity on piercing/sucking insects and is pending for registration.

Mode of Action

Avermectins block the neurotransmitter (g-aminobutyric acid (GABA) at the neuromuscular junction in insects and mites. Visible activity, such as feeding and egg laying, stops shortly after exposure, though death may not occur for several days.

FUMIGANTS

The fumigants are small, volatile, organic molecules that become gases at temperatures above 40°F. They are usually heavier than air and commonly contain one or more of the halogens (Cl, Br, or F). Most are highly penetrating, reaching through large masses of material. They are used to kill insects, insect eggs, nematodes, and certain microorganisms in buildings, warehouses, grain elevators, soils, and greenhouses and in packaged products such as dried fruits, beans, grain, and breakfast cereals.

Although its use is now in decline because of environmental concerns, methyl bromide is the most heavily used of the fumigants, 68,424 metric tons worldwide in 1996, almost half of which is used in the U.S.. The dominant use is for preplanting soil treatments, which accounted for 70% of that global total. Quarantine uses account for 5-8%, while 8% is used to treat perishable products, such as flowers and fruits, and 12% for nonperishable products, like nuts and timber. Approximately 6% is used for structural applications, as in drywood termite fumigation of infested buildings.

With the recently passed change to the Clean Air Act amendments of 1990, U.S. production and importation must be reduced 25% from 1991 levels by 1999. A 50% reduction must be achieved by 2001, followed by a 70% reduction in 2003, and a full ban of the product in 2005. Under the Montreal Protocol, developing countries have until 2015 to phase out methyl bromide production.

Some of the other common fumigants are ethylene dichloride, hydrogen cyanide, sulfuryl fluoride (Vikane®), Vapam®, Telone®II, D-D®, chlorothene, ethylene oxide, and the familiar home-use moth repellents napthalene crystals and paradichlorobenzene crystals.

Phosphine gas (PH_3) has also replaced methyl bromide in a few applications, primarily for insect pests of grain and food commodities. Treatment requires the use of aluminum or magnesium phosphide pellets, which react with atmospheric moisture to produce the gas. Phosphine, however, is very damaging to fresh commodities and is highly adsorbed onto oil, thus does not perform as a soil fumigant.

Alternatives that can fully replace methyl bromide are unlikely to be available by the deadlines set for replacement. Its low cost and utility on a wide variety of pests are hard to match. Because the loss of MeBr has considerable economic consequences, EPA has made it a priority to find and register replacements. To this end some progress has been made. The chemical 1,3-dichloropropane (Telone®) was registered in 2001 for preplant soil fumigation in strawberries and tomatoes. Moreover, iodomethane (Midas®) and metam-potassium (Curtin®) are both being evaluated as soil fumigants.

Mode of Action

Fumigants, as a group, are narcotics. That is, they act through means more physical than physiological. The fumigants are liposoluble (fat soluble); they have common symptomology; their effects are reversible; and their activity is altered very little by structural changes in their molecules. As narcotics, they induce narcosis, sleep, or unconsciousness, which in effect is their action on insects. Liposolubility appears to be an important factor in their action, since these narcotics lodge in lipid-containing tissues found throughout the insect body, including their nervous system.

INSECT REPELLENTS

Historically, repellents have included smoke, plants hung in dwellings or rubbed on the skin as the fresh plant or its brews, oils, pitches, tars, and various earths applied to the body. Before a more edified approach to insect olfaction and behavior was developed, it was wrongly assumed that if a substance was repugnant to humans it would likewise be repellent to annoying insects.

In recent history, the repellents have been dimethyl phthalate, Indalone®, Rutgers 612®, dibutyl phthalate, various MGK® repellents, benzyl benzoate, the military clothing repellent (N-butyl acetanilide), dimethyl carbate (Dimelone®) and diethyl toluamide (DEET, Delphene®). Of these, only DEET has survived, and is used worldwide for biting flies and mosquitoes. Most of the others have lost their registrations and are no longer available.

In 1999, EPA has registered a new insect repellent, N-methylneodecanamide. Rather than being used on humans to repel insects, it is applied to household floors and other surfaces to repel cockroaches and ants.

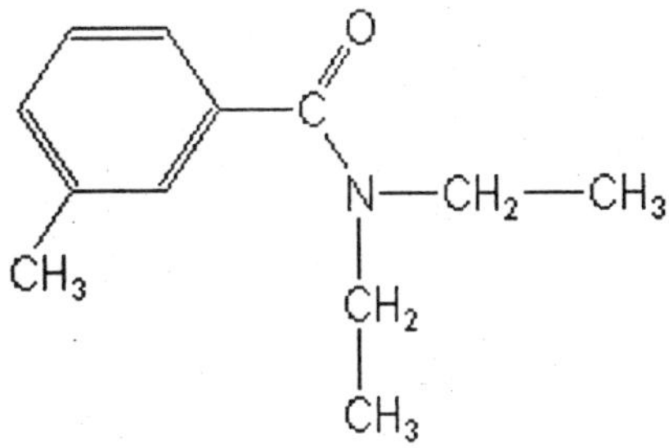

INORGANICS

Inorganic insecticides are those that do not contain carbon. Usually they are white crystals in their natural state, resembling the salts. They are stable chemicals, do not evaporate, and are usually water soluble.

Sulfur, mentioned in the introduction, is very likely the oldest known, effective insecticide. Sulfur and sulfur candles were burned by our great-grandparents for every conceivable purpose, from bedbug fumigation to the cleansing of a house just removed from quarantine of smallpox. Today, sulfur is a highly useful material in integrated pest management programs where target pests specificity is important. Sulfur dusts are especially toxic to mites of every variety, such as chiggers and spider mites, and to thrips and newly-hatched scale insects. Sulfur dusts and sprays are also fungicidal, particularly against powdery mildews.

Several other inorganic compounds have been used as insecticides: mercury, boron, thallium, arsenic, antimony, selenium, and fluoride. Arsenicals have included the copper arsenate, Paris green, lead arsenate, and calcium arsenate. The arsenicals uncouple oxidative phosphorylation, inhibit certain enzymes that contain sulfhydryl (-SH) groups, and coagulate protein by causing the shape or configuration of proteins to change.

The inorganic fluorides were sodium fluoride, barium fluosilicate, sodium silicofluoride, and cryolite (Kryocide®). Cryolite has returned in recent years as a relatively safe fruit and vegetable insecticide, used in integrated pest management programs. The fluoride ion inhibits many enzymes that contain iron, calcium, and magnesium. Several of these enzymes are involved in energy production in cells, as in the case of phosphatases and phosphorylases.

Boric acid, used against cockroaches and other crawling household pests in the 1930's and '40's, has also returned. As a salt, it is non-volatile and will remain effective as long as it is kept dry and in adequate concentration. Consequently, it has the longest residual activity of any insecticide used for crawling household insects, and is quite useful in the control of all cockroach species when placed in wall voids and other protected, difficult-to-reach sites. It acts as a stomach poison and insect cuticle wax absorber.

Sodium borate (disodium octaborate tetrahydrate) (Tim-Bor®, Bora-Care®) resembles boric acid in its action. This water-soluble salt is used to treat lumber and other wood products to control decay fungi, termites, and other wood infesting pests.

The last group of inorganics is the silica gels or silica aerogels—light, white, fluffy, silicate dusts used for household insect control. The silica aerogels kill insects by absorbing waxes from the insect cuticle, permitting the continuous loss of water from the insect body, causing the insects to become desiccated and die from dehydration. These include Dri-Die®, Drianone®, and Silikil Microcel®. Drianone® is fortified with pyrethrum and synergists to enhance its effectiveness.

REFERENCES

Palmer, WE, Bromley, PT, and Brandenburg, RL. *Wildlife & pesticides - Peanuts,* North Carolina Cooperative Extension Service. Retrieved on 2007.

Wells M., "Vanishing bees threaten US crops" (in English), *www.bbc.co.uk, BBC News.* Retrieved on 2007.

9

Characteristics of Herbicides

Herbicides, chemicals designed specifically to kill weeds, account for nearly 60% of all pesticide sales in the U.S. U.S. farmers spent an estimated $5.63 billion in 1998 for herbicides, with about $1.1 billion more spent in application costs. Herbicide use declined in 1999 to only a little more than $5.0 billion.

Weeds are objectionable to humans primarily because they reduce the quality and quantity of agricultural production, and produce allergens or contact dermatitis that affect public health. Approximately 10% of all plant species are weeds, or a total of some 30,000 weed species. Of these, 1,800 cause serious economic losses in crop production, and about 300 species plague cultivated crops throughout the world. The U.S. has become home to 70% of the world's worst weeds.

Herbicides, or chemical weed killers, have largely replaced mechanical methods of weed control in countries where intensive and highly mechanized agriculture is practiced. Herbicides provide a more effective and economical means of weed control than cultivation, hoeing, and hand pulling. Together with fertilizers, other pesticides, and improved plant varieties, they have made an important contribution to the increased yields we now have and serve to combat rising costs and shortages of agricultural labor. The heavy use of herbicides, although growing in many developing countries, is largely confined to North America, Western Europe, Latin America, Japan, and Australia. Without the use of herbicides, it would have been impossible to mechanize fully the production of cotton, sugar beets, grains, potatoes, and corn.

Herbicides are also used extensively away from the farm, in areas such as industrial sites, roadsides, ditch banks, irrigation canals, fence lines, recreational areas, lawns, railroad embankments, and power line rights-of-way. Herbicides remove undesirable plants that might cause damage, present fire hazards, or impede work crews. They also reduce costs of labor for mowing.

Herbicides are classed as *selective* when they are used to kill weeds without harming the crop and as *nonselective* when the purpose is to kill all vegetation. Both selective and nonselective materials can be applied to weed foliage or to soil containing weed seeds and seedlings, depending on the mode of action, and intended use pattern. The term *true selectivity* refers to the capacity of an herbicide, when applied at the proper dosage and time, to be active only against certain species of plants but not against others. But selectivity can also be achieved by placement, as when a nonselective herbicide is applied such that it reaches the weeds but not the crop.

Herbicide classification would be a simple if only the selective and nonselective categories existed. However, there are multiple classification schemes that may be based on selectivity, contact versus translocation, timing, area covered, and chemical classification.

Another scheme classifies herbicides into either contact or translocated categories. Contact herbicides kill the plant parts onto whose surfaces the chemical is applied and are most effective against annuals, those weeds that germinate from seeds and grow to maturity each year. Complete coverage is essential in weed control with contact materials.

Translocated herbicides are absorbed either by roots or above-ground parts of plants and are circulated within the plant system to distant tissues. Translocated herbicides may be effective against all weed types; however, their greatest advantage is in the control of established perennials, those weeds that continue their growth from year to year. Uniform application is a key requirement in achieving good weed control with translocated herbicides.

Herbicides are also classed according to timing of application, i.e., growth stage of crop or weed development. What constitutes the best timing varies by chemical class of the material and its persistence, the crop and its tolerance to the herbicide, weed species, cultural practices, climate, and soil type and condition. The three primary categories of application timing are preplanting, preemergence, and post emergence.

Preplanting applications for control of annual weeds are made to an area before the crop is planted, within a few days or weeks of planting and often includes a step to incorporate the herbicide in the soil. In recent years such preplant uses have included making fall applications of herbicides to suppress early weed growth the next spring before corn is planted. Preemergence applications are completed prior to emergence of the crop or weeds, depending on definition, after planting. Postemergence applications are made after the crop or weed emerges from the soil.

Herbicide application based on the area covered involves four categories: band, broadcast, spot treatments, and directed spraying. A band application treats a continuous strip, as along or in a crop row. Broadcast application covers the entire area, including the crop. Spot treatments are confined to small areas of weeds. Directed sprays are applied to crop stems, selected weeds or to the soil to avoid contact with crop foliage.

Over-the-top-application occurs when herbicides are applied over-the-top on the crop and weeds shortly after germination. The crops in these instances are naturally tolerant to the specific herbicide or have been genetically engineered to be tolerant. Examples of crops bioengineered to sustain broad spectrum herbicides include new varieties of cotton, corn, soybeans, canola, rice, sugarbeets, wheat and turf.

INORGANIC HERBICIDES

The first chemicals used in weed control were inorganic compounds. Brine and a mixture of salt and ashes were both used by the Romans to sterilize the land as early as biblical time. In 1896, copper sulfate was used selectively to kill weeds in grain fields. From about 1906 until 1960, sodium arsenite solutions were the standard herbicide of commerce. Arsenic trioxide has been used at rates of 400 to 800 lbs/acre for soil sterilization, whereas 1 or 2 lbs of currently used, selected organic herbicides would easily produce the same result. The trivalent arsenicals are nonspecific inhibitors of those enzymes containing sulfhydryl groups, and also uncouple oxidative phosphorylation.

Ammonium sulfamate ($NH_4SO_3NH_2$), another salt, was introduced in 1942 for brush control. Other salts have been used over the years. The borate herbicides (sodium tetraborate, sodium borate, etc.) were used heavily for soil sterilization. A nonselective herbicide used extensively for the last 50 years is sodium chlorate ($NaClO_3$). It acts as a soil sterilant at rates of 200 lbs/acre, but it can be used as a foliar spray at 5 lbs/acre as a cotton defoliant. Several of the inorganics are still useful in weed and brush control, but are rapidly being replaced by organic materials. EPA has placed heavy restrictions on some inorganics because of their persistence in soils. Advice to the reader: inorganics are not wise choices for use around the home except by experts intending to remove all vegetation from an area.

ORGANIC HERBICIDES

The categorization of organics is rather complex, comprising some 31 classes, if the old, now illegal, category of petroleum oils is included.

Chloroacetamides

The chloroacetamides and amides are both meristematic growth inhibitors. They differ in that the amides are not translocated.

The chloroacetamides are used as preemergence herbicides, some preplant soil incorporated and others surface applied. Of these, alachlor was one of the first to gain wide acceptance. In 1999, it was the 17th most used pesticide, in the nation with an estimated 7-10 million pounds being used in agriculture, mainly on corn, cotton and soybeans.

Metolachlor (Dual Magnum®) is soil-incorporated as a preplant and preemergence herbicide for annual grass and some broadleaf weeds in corn, soybeans and peanuts. Metolachlor and a more recent resolved isomer form (s-metolachlor) together accounted for a total use of 42-50 million pounds in 1999 Acetochlor (Harness®, Surpass®, TopNotch®), not illustrated, was first registered for crop use in 1994. In 1999, more than 30 million pounds of acetochlor was used in US agriculture, mostly on corn. Butachlor is used only on rice, but not in the U.S., to control grass weeds and is applied either preplant incorporated or preemergence to seeded or transplant rice. Propachlor is used on corn, cotton, sorghum and soybeans to control mostly grass weeds. It is applied to the soil surface prior to weed and crop emergence. One new member of this class has surfaced recently. It is thenylchlor (Kasamets®) which is slated for use on rice in South Korea.

METIKACHLOR (DUAL MAGNUM®)

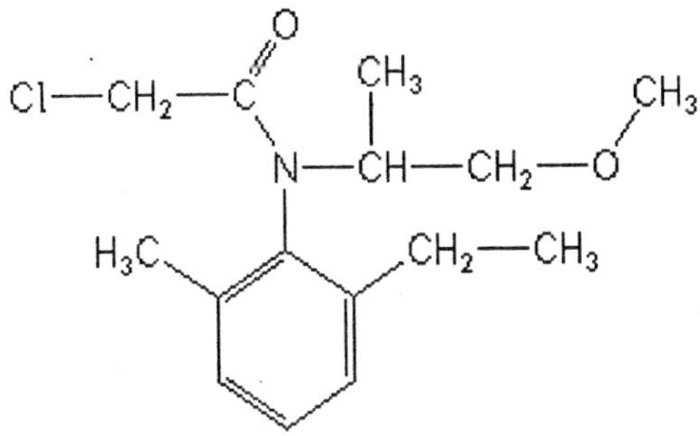

2-chloro-N-(2-ethyl-6-methylphenyl)-N-(2-methoxy-1-methylethyl)acetamide.

The mode of action for this class is predominantly inhibition of very long chain fatty acids.

Amides

The amide herbicides have diverse biological properties. Amides are sometimes grouped with the chloroacetamides. Dimethenamid (Frontier®), first registered in 1993, controls the major grass weeds, broadleafs and yellow nutsedge in corn, soybeans, peanuts, dry beans and sorghum as a selective preemergence herbicide. A resolved form, in which the product was purified to yield the active s isomer, was developed and registered in 1999. This product is named dimethenamid-P (Outlook®).

Propanil (Stampede®) has been used extensively on rice fields as a selective postemergence control for a broad spectrum of weeds. Napropamide is used in the control of grass and broadleaf weeds in vineyards and orchards, and in direct-seeded tomatoes, strawberries, ornamentals, and tobacco.

PROPANIL (STAMPEDE®)

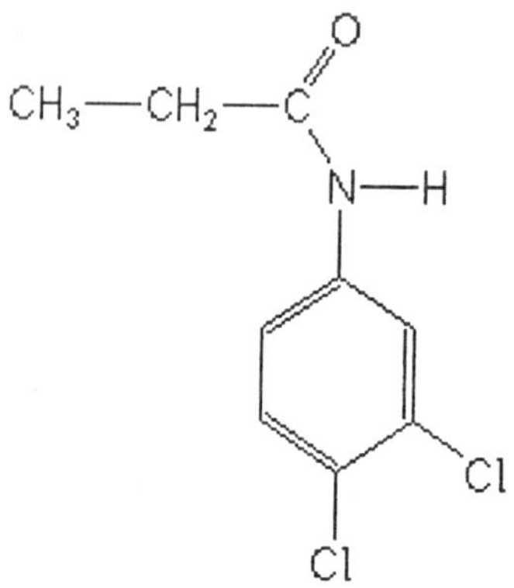

N-*(3,4-dichlorophenyl) propanamide*

Propanil acts primarily in the leaves and is a strong inhibitor of the *Hill reaction*. This is a light-initiated reaction that splits water (photolysis), resulting in the production of free oxygen (O_2) by plants. Chlorophyll, the green pigment of plants, is an essential ingredient in the reaction, since it catalyzes the production of oxygen from water and the transfer of the hydrogen to a hydrogen-acceptor.

Naptaḷam is used as preemergence sprays for controlling germinating seedlings of both grasses and broadleaf weeds. Its mode of action is the blocking of the growth control regulator ,indoleacetic acid. Pronamide is used as a selective, preemergence and early postemergence herbicide, for both grass and broadleaf weeds in a broad range of crops. It is taken up by roots, and has little foliar activity. Bensulide is one of the better turf herbicides, especially for the control of crabgrass. It could be classified as an organophosphate herbicide, and has higher mammalian toxicity than most herbicides because it is an organophosphate. Bensulide acts by inhibiting cell division in root tips. It is used as a preemergence herbicide in lawns to control certain grasses and broadleaf weeds, but it fails as a foliar spray because it is not translocated

A new member of this class is pethoxamid (Korban®, Successor®), destined for use in Europe for weed control in corn and soybeans. This product, like dimethenamid, is an inhibitor of very long chain fatty acids.

There is no consistent pattern in the mechanism of action of amide herbicides. Some are applied only to the soil and are active through the root system or seeds, while others are applied only to foliage.

Organoarsenicals

The organic arsenicals are no longer widely used. The class includes derivatives of arsinic and arsonic acids. Cacodylic acid (dimethylarsinic acid) and its sodium salt are the only derivatives of arsinic acid. Disodium methanearsonate (DSMA) and monosodium methanearsonate (MSMA) are salts of arsonic acid.

MSMA

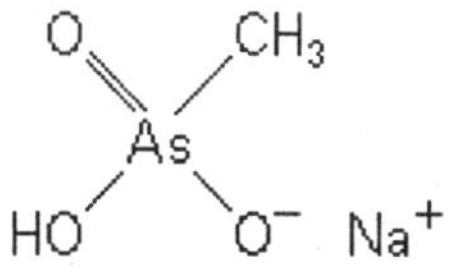

Monosodium methylarsonate

The organic arsenicals are much less toxic to mammals than the inorganic forms, are crystalline solids, and are relatively soluble in water. The arsonate, or pentavalent arsenic, acts in a way different from the trivalent form of the inorganic arsenicals just described. The arsonates upset plant metabolism and interfere with normal growth by entering into reactions in place of phosphate. They not only substitute for essential phosphate but also are absorbed and translocated in a manner similar to that of phosphates.

The trivalent arsenates are exclusively contact herbicides, whereas the pentavalent arsonates are translocated to underground tubers and rhizomes, making them extremely useful against johnsongrass and nut sedges. They are often applied as spot treatments.

Benzoic Acids

Several derivatives of benzoic acid are employed as herbicides and are applied to the soil against germinating seeds and seedlings. They are classed as growth hormone herbicides, and move both from leaves to the terminal meristems of leaf, shoot and root, and also move in the transpiration stream, permitting them to also be soil applied. Their mode of action is thought to be the interference in protein synthesis with the resulting interference in cell formation. Dicamba has been one of the 25 most used pesticides in the U.S. for nearly two decades. Recent annual usage rates have been 6-8 million pounds. Chlorfenac and Chloramben have been discontinued.

DICAMBA (BANVEL®)

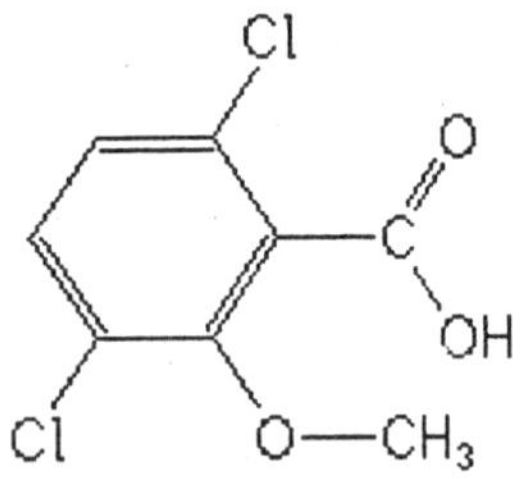

3,6-dichloro-2-methoxybenzoic acid

Nitriles

Nitriles have a benzene ring, containing the CN or cyanide

grouping. These two compounds, dichlobenil and bromoxynil, are contact herbicides which control broadleaf weeds in grass-type crops. They are registered for use on small grains, corn, grain sorghum, and for seedling turf for postemergence control of broadleaf weeds.

DICHLOBENIL (CASORON®)

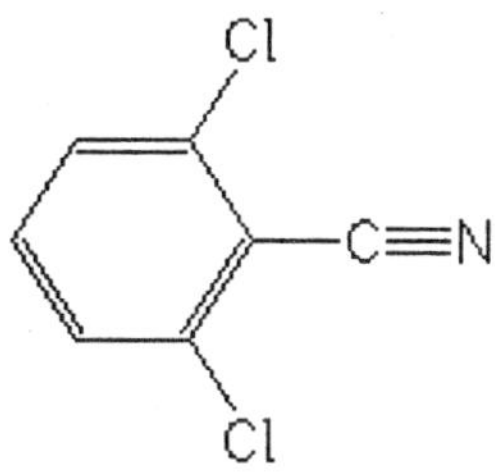

2,6-dichlorobenzonitrile

Their mechanisms of action are broad, involving seedling growth inhibition, potato sprout inhibition, and gross disruption of tissues by inhibiting oxidative phosphorylation and preventing the fixation of CO_2. These effects, however, do not explain their rapid action. Ioxynil belongs to the nitriles but is not registered in the U.S.

Benzothiadiazoles

The single representative of this group is bentazone. Bentazone kills broadleaf weeds and nutsedge by inhibiting photosynthesis. It differs from other photosynthesis-inhibitors, such as the triazines and phenylureas, in that very limited translocation occurs. Consequently uses are limited to postemergence treatments and require thorough plant coverage to be effective. Bentazon is effective in controlling broadleaf weeds in soybeans, rice, corn, peanuts, beans, and peas.

BENTAZONE (BASAGRAN®)

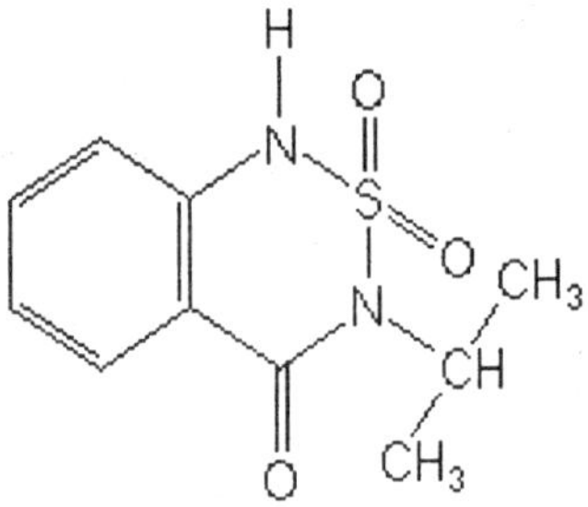

3-(1-methylethyl)-1H-2,1,3-benzothiadiazin-4(3H)-one 2,2-dioxide

Bipyridyliums

The name *bipyridylium* suggests the attachment of two pyridyl rings. There are three important herbicides in this group, diquat, paraquat and difenzoquat. All are contact herbicides that damage plant tissues quickly, causing the plants to appear frostbitten because of cell membrane destruction. This rapid wilting and desiccation occur within hours, making these novel herbicides also useful as preharvest desiccants for seed crops, cotton, soybeans, sugarcane, and sunflowers. Diquat is also used in aquatic weed control. None are active in soils.

The action of these herbicides occurs through positive ions, naturally dissociated, which are reduced by photosynthesis to form stable free radicals. These free radicals are then oxidized to reform the original ion and hydrogen peroxide which destroys plant tissue.

PARAQUAT

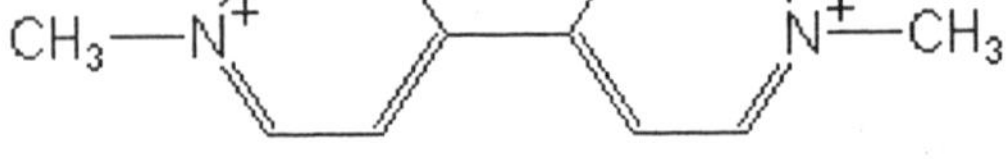

1,1′-dimethyl-4,4′-bipyridinium

Carbamates

The carbamates are esters of carbamic acid, and some representatives are active as herbicides, and some as fungicides or insecticides. Discovered in 1945, the carbamates described here are used primarily as selective preemergence herbicides, but some are also effective in postemergence use.

The first of the herbicidal carbamates was propham (also known as IPC); it was followed by chlorpropham and later by asulam. These herbicides kill plants by stopping cell division and plant tissue growth. Two effects are noted: cessation of protein production and shortening of chromosomes undergoing mitosis (duplication). Carbamates are readily translocated and inhibit meristematic development.

PROPHAM (TURBERITE®)

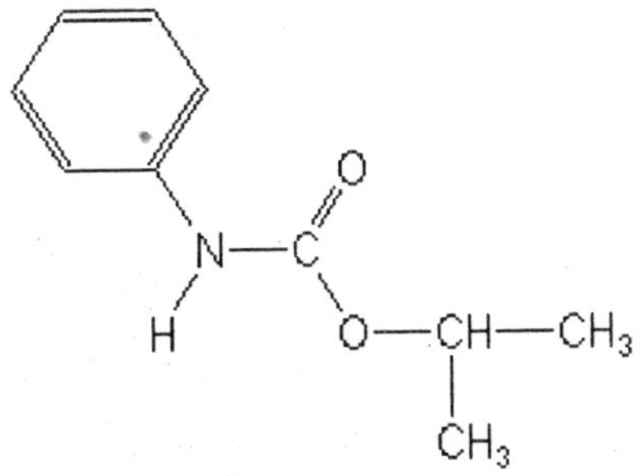

1-methylethyl phenylcarbamate

Phenyl-carbamates

Similar to the carbamates are the phenyl-carbamates, of which there are but two. Phenmedipham contains two carbamate radicals in its molecule and is used as a postemergence herbicide in sugar beet & sunflower production. Desmedipham (Betanex®) is registered only for sugar beets in the U.S., for a wide range of broadleaf spring weeds.

PHENMEDIPHAM (BETANAL®)

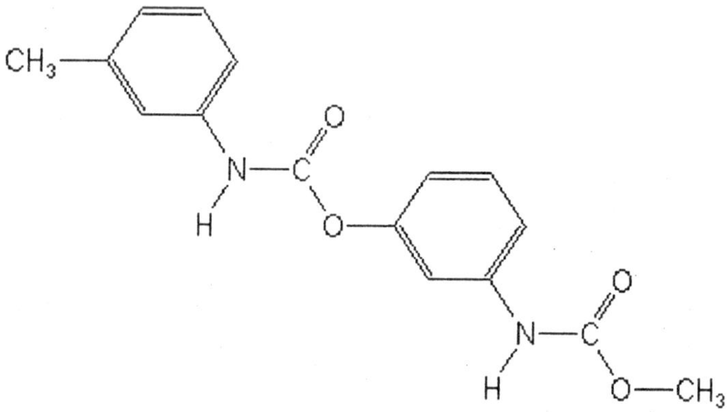

3-[(methoxycarbonyl)amino]phenyl (3-methylphenyl) carbamate

Chlorinated Aliphatic Acids

Two aliphatic acids heavily used as herbicides were TCA and dalapon, used against grasses, particularly quackgrass and Bermudagrass. They both act by precipitation of protein within the cells. Dalapon was widely used around homes to control Bermudagrass. Neither product is marketed in the U.S. today.

Cyclohexanediones

The cyclohexanediones are all selective, postemergence herbicides used for the control of annual and perennial grass weeds in broadleaf crops, e.g., cotton, soybeans, potatoes, sugarbeets, and vegetables. They do not control broadleaf weeds or sedges. Only two materials are registered in the U.S., sethoxydim, and clethodim.

Cycloxydim and tralkoxydim are not registered but one new member of the class has recently made an appearance. It is tepraloxydim (Aramo®) which controls various grass weeds in cotton, onions, soybeans and sugarbeets. Although these herbicides do interfere with auxins their primary action is to inhibit lipid biosynthesis.

SETHOXYDIM (POAST®, VANTAGE®)

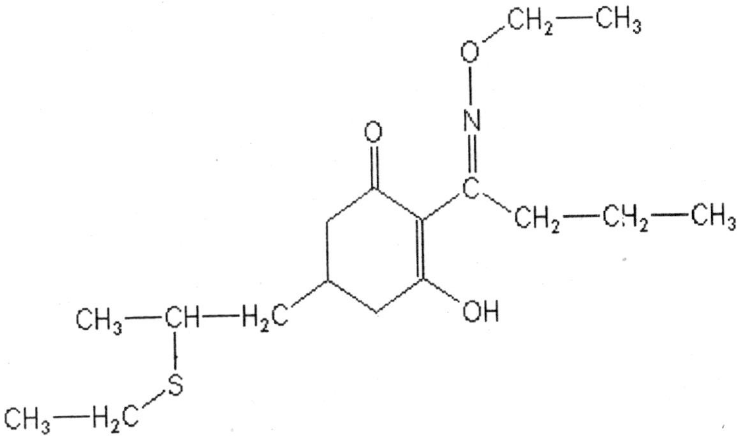

2-[1-(ethoxyimino)butyl]-5-[2-(ethylthio)propyl]-3-hydroxy-2-cyclohexen-1-one

Dinitroanilines

The dinitroanilines are some of the most heavily used in agriculture, and are used almost exclusively as soil-incorporated, preemergence selective herbicides in many field crops. Examples of substituted dinitroanilines are trifluralin, benefin, oryzalin, pendimethalin, and isopropalin. Trifluralin has very low water solubility, which minimizes leaching and movement from the target. The nitroanilines inhibit both root and shoot growth when absorbed by roots, but they have an involved mode of action, which includes inhibiting the development of several enzymes and the uncoupling of oxidative phosphorylation.

TRIFLURALIN(TREFLAN®)

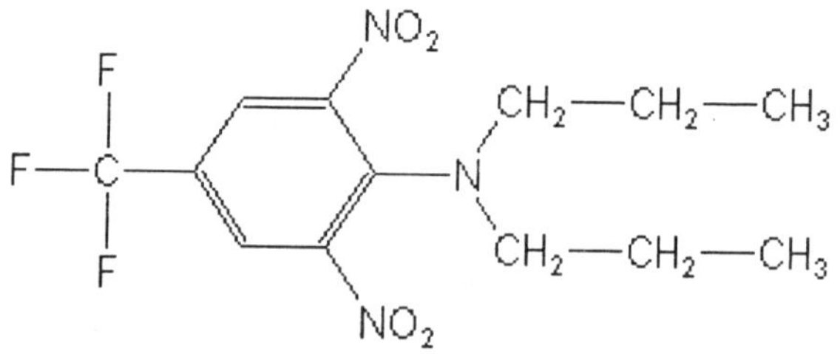

*2,6-dinitro-*N,N*-dipropyl-4-(trifluoromethyl) benzenamine*

Oryzalin controls annual grasses and broadleaf weeds in rice, cotton, soybeans, nonbearing fruit trees, nut trees, vineyards, and ornamentals. Pendimethalin controls annual grass and certain broadleaf weeds in corn and is preplant incorporated in cotton, soybeans, vegetables, and tobacco. As preemergence herbicides, the dinitroanilines severely inhibit growth of seedling roots, resulting in plant stunting. Prodiamine (Barricade®) is used for turf and ornamentals and Ethalfluralin (Sonalan®), a newer nitroaniline, for cotton and soybeans. It is soil-incorporated selective for annual grasses and broadleaf weeds, and especially useful against nightshade.

Dinitrophenols

The dinitrophenols included DNOC, DINOSEB (Dinitro®, DNBP), and some lesser known materials. These were used for more than 50 years as herbicides, ovicides, fungicides, insecticides, and blossom-thinners. The dinitrophenols act by uncoupling oxidative phosphorylation, not only in plants, but also in warm blooded animals, including man. Consequently, by general agreement with the manufacturers, EPA cancelled all uses of this herbicide class in 1987, because of suspected long term health effects in man.

DINOSEB

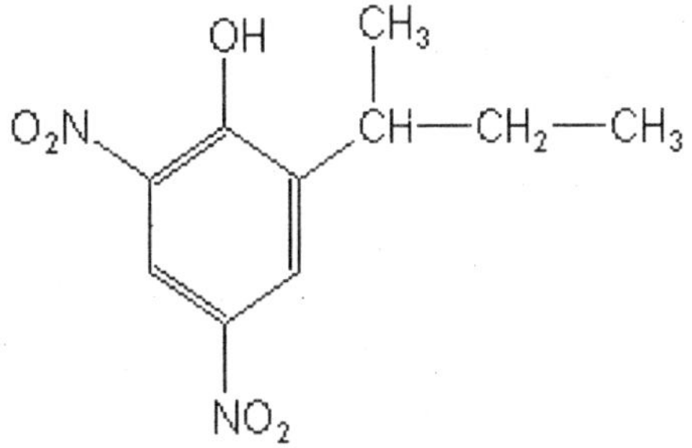

2-(1-methylpropyl)-4,6-dinitrophenol

Diphenyl Ethers

The diphenyl ether herbicides are constructed around a molecule made of two phenyl (benzene) rings tied together with oxygen.

The first of this group was nitrofen, introduced in 1963. They are generally used in preemergence or early postemergence weed control. They are primarily contact herbicides and are readily absorbed by leaves and roots. Translocation is limited. They induce chlorosis and necrosis as visible effects of activity. Their mode of action is thought to be inhibition of electron transport and coupled photophosphorylation. In some instances they induce growth responses, resembling the action of auxins and the phenoxy herbicides.

Some of the more successful herbicides are found in this group. They are foliar applied, do not translocate, and are enhanced by high volume sprays containing surfactants to increase coverage. They are used to control broadleaf and grass weeds in soybeans, peanuts, and other legumes, postemergence weed control in wheat, barley, soybeans and sugar beets. They include acifluorfen, lactofen, and oxyfluorfen. Two others are bifenox (Modown®), and fomesafen (Reflex®, Tornado®).

ACIFLUOFEN (BLAZER®)

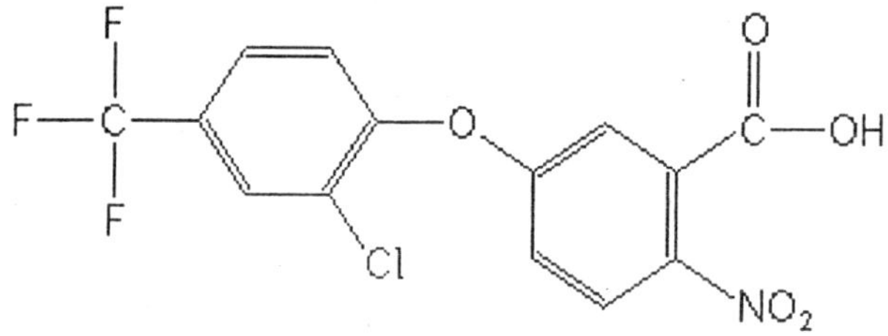

5-[2-chloro-4-(trifluoromethyl)phenoxy]-2-nitrobenzoic acid

Imidazolinones

The imidazolinones are a relatively new class, first appearing in 1981.The most recent imazamox (Raptor®) was introduced in 1998. They are meristematic inhibitors which inhibit amino acid branched-chain biosynthesis. They are more effective against broadleaf weeds than grasses, and have both foliar and soil activity. Soil persistence ranges from moderate to long.

Imazapyr has been used the longest, as a nonselective, broad-spectrum herbicide with residual activity in noncrop areas and forestry. Imazaquin is a selective pre- and postemergence herbicide which has enjoyed wide use on soybeans. Imazamethabenz-methyl is a selective postemergence herbicide against wild oats, mustards and buckwheat in sunflowers, wheat and barley. Imazethapyr is a preemergence, preplant incorporated and postemergence material for control of annual and perennial grass and broadleaf weeds in soybeans and other legumes.

IMAZAPYAR (ARSENAL®)

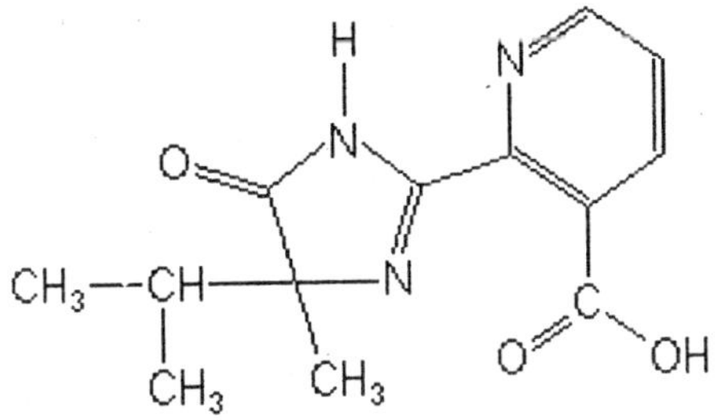

*2-[4,5-dihydro-4-methyl-4-(1-methylethyl)-5-oxo-1*H*-imidazol-2-yl]-3-pyridinecarboxylic acid*

Triazolopyrimidines

This class, closely related to the imidazolinones, were introduced by Dow AgroSciences in 1990. Flumetsulam (Broadstrike®) was the first to be introduced, followed by metosulam (Eclipse®), chloransulam (Firstrate®), and diclosulam (Strongarm®). Flumetsulam is a selective preplant, preemergence and postemergence material for tough broadleaf weeds in corn and soybeans, and sold in the U.S. only as mixtures with other herbicides. Metosulam, similar in selectivity, is not registered in the U.S. Chloransulam and diclosulam are being developed in the U.S. for broadleaf weed control on soybeans and peanuts. New entrants to the class are penoxsulam and florasulam (Primus®); the latter is in development to control broadleaf weeds in corn and cereals and the former is slated for use on rice. Their mode of action is similar to that of the sulfonylureas and imidazolinones.

FLUMETSULAM (BROADSTRIKE®)

Aryloxphenoxy Propionates (formerly Oxyphenoxy Acid Esters)

This recently introduced class of herbicides is a family of

postemergents with high activity against all grasses at economical rates. Consequently, they are applied to most broadleaf crops with little risk of crop injury.

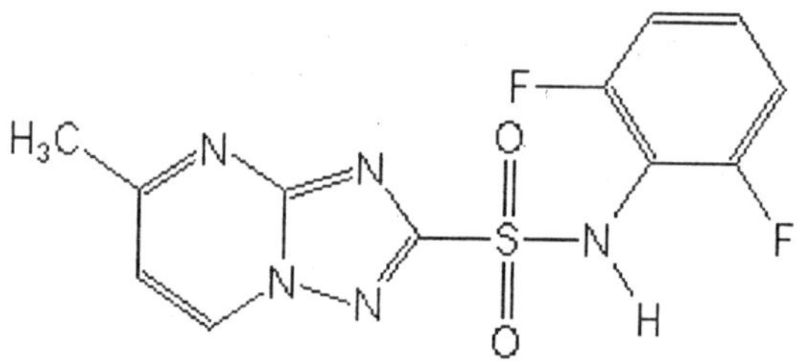

N-*(2,6-difluorophenyl)-5-methyl[1,2,4]triazolo[1,5-*a*]pyrimidine-2-sulfonamide*

At higher rates of application some soil activity is observed. In grasses the compounds are readily translocated from point of uptake to the growing meristem. Weather apparently has little effect on activity. Notably, they are all effective against the arch enemies crabgrass and johnsongrass. Fluazifop-butyl is the most widely used of the group, for postemergence control of both perennial and annual grass weeds, over-the-top in broadleaved crops, especially cotton and soybeans. Some recent additions to the class include propaquizafop, quizalofop-P, metamifop and pyriftalid.

FLUAZIPFOP-BUTYL (FUSILADE®)

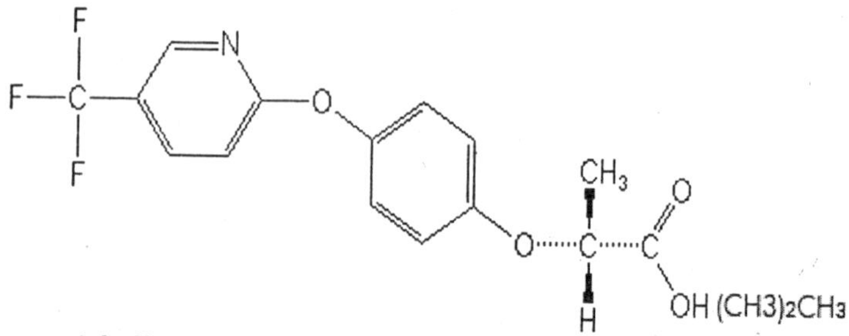

2-[4-[[5-(trifluoromethyl)-2-pyridinyl]oxy]phenoxy]propanoic acid

Phenoxys

2,4-D was introduced in 1944 as the first of the "phenoxy herbicides," "phenoxyacetic acid derivatives," or "hormone" weed killers. These were highly selective for broadleaf weeds and were translocated throughout the plant. 2,4-D provided most of the impetus in the commercial search for other organic herbicides in the 1940s. Several compounds belong to this group, of which 2,4-D and 2,4,5-T are the most familiar. Other important compounds in this group are 2,4-DB, MCPA, and silvex.

2,4-D

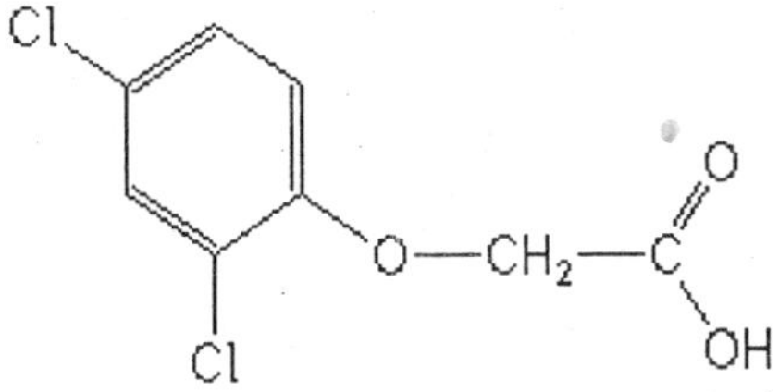

2,4-dichlorophenoxy)acetic acid

The phenoxy herbicides have complex mechanisms of action resembling those of auxins (growth hormones). They affect cellular division, activate phosphate metabolism, and modify nucleic acid metabolism.

2,4-D, MCPA, and 2,4,5-T have been used for years in very large volume worldwide. The latter product, 2,4,5-T, used mainly to control woody perennials, became the subject of extended investigation, particularly because of its use in Vietnam in combination with 2,4-D as Agent Orange. Certain samples were found to contain small amounts of a highly toxic impurity, 2,3,7,8-tetrachlorodibenzo-*p*-dioxin, commonly referred to as tetrachlorodioxin, or dioxin. Although alterations in manufacturing procedures reduced the dioxin content to minimal levels, 2,4,5-T registrations were cancelled and the product voluntarily removed by the manufacturers in 1985.

Agent Orange was replaced with Agent White, a mixture of 2,4-D and picloram, which was longer lasting and more effective. That same combination is used today (Grazon®P+D, Tordon®RTU) on pastures and rangeland for control of woody species.

2,4,-D has been and continues to be one of the most useful herbicides developed. More than 33 million pounds manufactured in the United States, is used each year in 35 ester and salt forms. In agriculture it is used on cereal, grain crops and sugar cane for the control of broadleaf weeds, and on rights-of-way, turf and lawns, and in forest conservation programs. The manufacturing process for 2,4-D used in the United States does not result in any level of tetrachlorodioxin contamination. Other members of the phenoxys in wide use are dichlorprop (2,4-DP), the butyric acid form (MCPB), and mecoprop (MCPP).

Ureas

Thousands of ureas, also called substituted ureas, have been tested as herbicides and many are in use today. Fluometuron is illustrated. Others include linuron, diuron and monuron, fenuron-TCA, siduron and tebuthiuron. Most of the ureas are relatively nonselective and are usually applied to the soil as preemergence herbicides; some have postemergence uses, while others are active when foliar applied. The ureas are strongly absorbed by the soil, then absorbed by roots. Their mechanism of action is to inhibit photosynthesis—the production of plant sugars—and, indirectly, to inhibit the Hill reaction. Despite the number and age of this group new ones continue to make an appearance. Two recent additions include isoproturon and cumyluron, the latter of which though structurally a urea has an unknown mode of action. Neither is used in the U.S. There are yet a larger number of ureas not registered for use in the U.S.

FLUMETURON (COTORAN®)

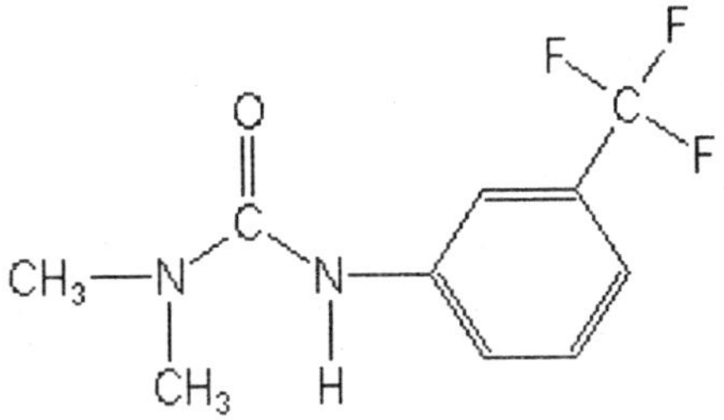

N,N-*dimethyl*-N′-*[3-(trifluoromethyl)phenyl]urea*

Phosphono Amino Acids

These are foliar applied, translocated herbicides that interfere with normal plant amino acid synthesis. They are more effective against grasses than broadleaf weeds, and are considered to be nonselective foliar compounds with no soil activity. They penetrate rather slowly, thus rainfall shortly after application can reduce effectiveness. Found here are glyphosate, glufosinate, fosamine and glyphosate trimesium.

GLYPHOSATE (ROUNDUP®)

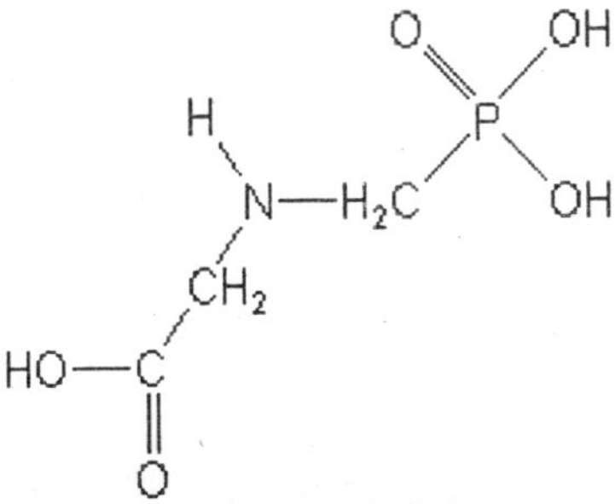

N-*(phosphonomethyl)glycine*

Glyphosate, now the most widely used pesticide in the U.S. and in the world is a nonselective, nonresidual, postemergence material. It is recognized for its effectiveness against perennial, deep-rooted, grass and broadleaf weeds, as well as woody brush problems in crop and noncrop areas. It is a translocated, foliar-applied herbicide that can be applied at any stage of plant growth or at any time of year, with most types of application equipment, including the new wick, roller, and wiper devices. Its mechanism of action appears to be the inhibition of the synthesis of aromatic amino acids, which results in the inhibition of nucleic acid metabolism and protein synthesis. There are a number of salt forms registered for glyphosate.

The growth of glyphosate was dramatically enhanced starting in the mid-1990s by the creation and marketing of several crops (corn, cotton, soybeans, canola with others planned) bioengineered to render them tolerant to applications of glyphosate. This technological advance allows farmers to spray the non-selective herbicide onto these crops to achieve excellent weed control without significant crop damage. Although similar traits were introduced into certain crops for other herbicides including imadazolinones and glufosinate these have not matched the commercial success enjoyed by glyphosate.

Glufosinate is used primarily for grass weeds in orchards and vineyards as a postemergence herbicide. Surprisingly, it is used as a desiccant on potatoes, and for weed control in minimum tillage systems. Fosamine controls bud growth in woody species in noncrop areas, and glyphosate trimesium (Touchdown®) is the newest of the lot, appearing in 1988.

Phthalic Acids

Only two compounds of significance fall into this minor class. They are chlorthal (DCPA; often sold as the dimethyl ether which is called chlorthal-dimethyl), developed in 1960, and endothall, which appeared in 1965. Chlorothal is a quality crabgrass herbicide, but is also highly effective against most grass and broadleaf weeds in vegetables. Endothall is used in sugar beets for a narrow range of weeds, as an aquatic herbicide, as an algicide, and as a desiccant for cotton, potatoes, alfalfa and clover. Its low toxicity to fish makes it an ideal aquatic herbicide, and an outstanding example of environmental protection through pesticide selectivity.

DCPA, CHLORTHAL (DACTHAL®)

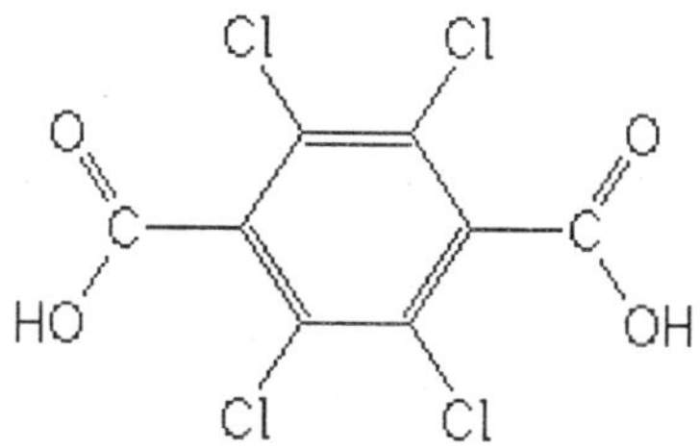

2,3,5,6-tetrachloro-1,4-benzenedicarboxylic acid

Pyridazinones

Because these herbicides inhibit the formation of chlorophyll, the green pigment in plants, the weeds soon lose their color following uptake, and thus appear bleached. They are also referred to as being carotinoid biosynthesis inhibitors. These newer herbicides control grasses and broadleaf weeds, are compatible with the bipyridiliums and a wide range of preplant soil incorporated preemergence herbicides.

Pyrazon (formerly chloridizon), used almost exclusively on sugar beets, is a selective, pre- and postemergence herbicide for broadleaf weeds and is unusual for the group in being a photosystem II inhibitor.

Norflurazon is used as a preemergence herbicide for control of annual grasses and certain broadleaf weeds in cotton, soybeans, and vine and deciduous fruit crops. Fluridone is an aquatic herbicide used in ponds, drainage and irrigation canals, lakes and reservoirs, for the control of most aquatic weeds at any stage of growth.

NORFLURAZON (SOLICAM®, EVITAL®)

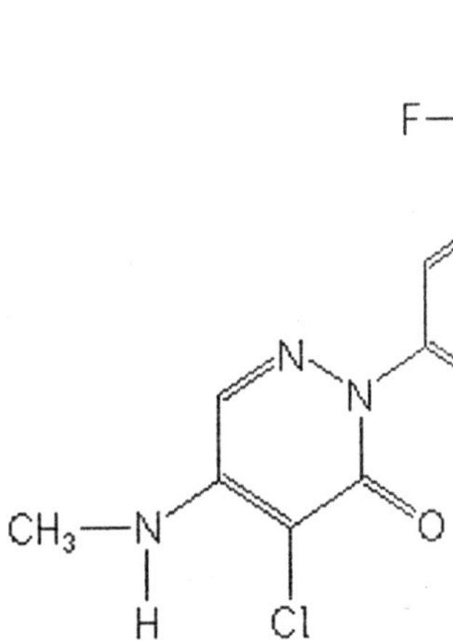

4-chloro-5-(methylamino)-2-[3-(trifluoromethyl)phenyl]-3(2H)-pyridazinone

Fluorochloridone (Racer®), an experimental pre-emergence herbicide for grass and broadleaf weeds in a wide range of crops. Clomazone (formerly dimethazone) is a pre-emergence or preplant incorporated herbicide for grass and broadleaf weeds in soybeans, cotton and vegetables.

Oxadiazon is used for weed control in dry-seeded rice, and for turf and ornamentals. Among the newest family members are beflubutamid, picolinafen and oxadiargyl. The first is an experiment being evaluated for use on small grains. Oxadiargyl is not used in the U.S. Oxadiargyl like oxadiazon (Ronstar®) is an inhibitor of protoporphyrinogen oxidase. Picolinafen (Pico®) has potential uses on small grains and is an inhibitor of phytoene desaturase.

Carboxylic Acids (formerly Pyridinoxy and Picolinic Acids)

The carboxylic acid herbicides like their benzoic acid relatives are identified by the carboxylic acid radical attached to the benzene ring. More specifically, this COOH group is usually referred to as the *carboxyl* group. In this group are picloram, triclopyr, fluoroxypyr, clopyralid and quinclorac. These are hormone herbicides that translocate in both the phloem and xylem. They are ideal for control of perennial broadleaf weeds and brush, and have both soil and foliar activity. Picloram is long-lived in soil, while triclopyr and clopyralid are much less persistent. Their mode of action is that of the synthetic auxins, acting like the natural plant hormone indoleacetic acid, and like their kindred phenoxy-, and benzoic-acid herbicides.

PICLORAM (TORDON®)

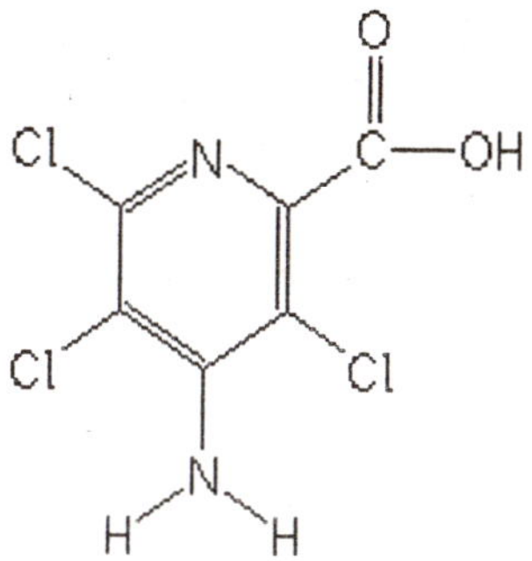

4-amino-3,5,6-trichloro-2-pyridinecarboxylic acid

Picloram, combined with 2,4-D to form Agent White, was used in Viet Nam as a replacement for Agent Orange (2,4-D and 2,4,5-T). That same combination is used today on pastures and rangeland for the control of woody species under the proprietary names of Grazon®P+D, Pathway®, and Tordon®RTU.

Pyridines

Pyridine is a benzene ring with one of the carbons replaced by a nitrogen, as in the illustration. There are several substituted pyridines used as herbicides. In this group are dithiopyr (Dimension®) and thiazopyr (Visor®). Dithiopyr is a selective pre- and postemergence material, used only on turf to control a wide variety of grass weeds. It is frequently formulated with other herbicides and on fertilizers. Thiazopyr is a selective preemergence compound that works well for practically all grass weeds, on a wide variety of crops, including citrus, cotton, corn, peanuts, soybeans and potatoes. They appear to be microtubule assembly inhibitors.

DITHIOPYR (DIMENSION®)

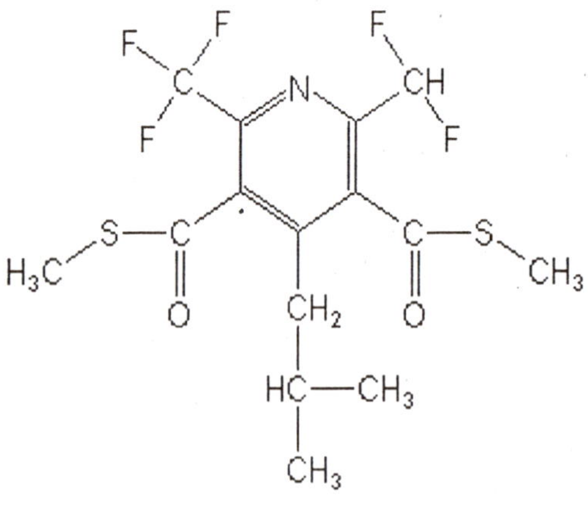

S,S′-dimethyl 2-(difluoromethyl)-4-(2-methylpropyl)-6-(trifluoromethyl)-3,5-pyridinedicarbothioate

Sulfonylureas

The sulfonylureas are meristematic inhibitors with both foliar and soil activity which control broadleaf weeds better than grasses. They have very specific activity with application rates often being less than one-fourth ounce (7 grams) per acre. They range in persistence from very long to moderate, and have very low mammalian toxicity. The mode of action for the sulfonylureas is the inhibition of branched chain amino acid biosynthesis through the inhibition of an enzyme called acetolactate synthetase (ALS).

In this group belong chlorsulfuron, chlorimuron-ethyl, metsulfuron-methyl, and sulfometuron-methyl. Chlorsulfuron appeared in 1981, and works well for broadleaf weeds in wheat, barley, oats and non-crop weed control. Wild oats are tolerant to this material. Chlorimuron-ethyl controls broadleaf weeds in soybeans as a postemergence herbicide. Metsulfuron-methyl is used in the same role as chlorsulfuron. Sulfometuron-methyl is used only on noncrop areas for nonselective weed control as a broad spectrum pre- and postemergence herbicide. It can also be used in pine forests to control woody tree species. Thifensulfuron-methyl is a selective postemergence material for broadleaf weeds and wild garlic in small grains. Halosulfuron-methyl (Battalion®, Permit®) is a selective pre- and postemergence compound for corn, sorghum, rice, sugarcane and turf.

CHLORSULFURON (GLEAN®)

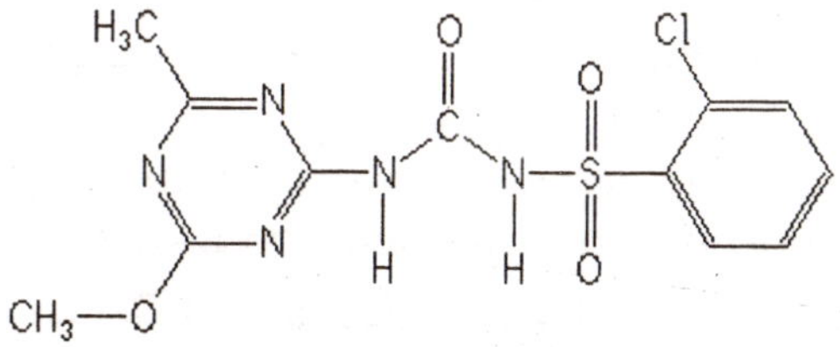

Additional older sulfonylureas are bensulfuron (Londax®), nicosulfuron (Accent®), triasulfuron (Amber®, Logran®) and primisulfuron-methyl (Beacon®, Tell®).

The newer sulfonylureas are amidosulfuron (Eagle®, Pursuit®), azimsulfuron (Gulliver®), cyclosulfamuron (Invest®), ethametsulfuron-methyl (Muster®), ethoxysulfuron (Sunrise®), flupyrsulfuron-methyl (Lexus®), foramsulfuron (Equip®, Option®), iodosulfuron-methyl (Husar®), mesosulfuron-methyl (Mesomaxx®), oxasulfuron (Dynam®, Expert®), prosulfuron (Peak®), rimsulfuron (Matrix®, Shadeout®), sulfosulfuron (Maverick®, Sundance®), tribenuron-methyl (Express®), trifloxysufuron (Brawn®, Enfield®), triflusulfuron-methyl (Debut®, Upbeet®), and tritosulfuron (Corto®). Foramsulfuron and Iodosulfuron-methyl were registered by the U.S. EPA in 2002 while trifloxysulfuron was registered in 2003. For the past two decades this has been among the most fertile groups for new herbicide discovery of all herbicide classes, and though now somewhat mature may continue to receive new candidates from the major U.S. and Japanese manufacturers.

Thiocarbamates

The thiocarbamates contains sulfur (from the Greek word *theion*) and are derived from the hypothetical thiocarbamic acid.

The thiocarbamates are selective herbicides marketed for weed control in croplands. The thio- and dithiocarbamates have exceptional vapor pressures, are quite volatile, and must be incorporated in the soil at time of application. They inhibit the development of seedling shoots and roots as they emerge from the weed seeds. Consequently, they are all used as either preplant or preemergence, soil-incorporated herbicides.

Examples of thiocarbamates are EPTC and pebulate. EPTC is used mostly to control annual and perennial grassy weeds, including nutgrass, johnsongrass, and quackgrass. Pebulate is effective against both grassy and broadleaf weeds in tobacco, and tomatoes. Also included among the thiocarbamates are thiobencarb, butylate, cycloate, molinate, and vernolate. Thiobencarb is the most widely used rice herbicide Worldwide. It is a preemergent and early postemergent compound, for grasses and broadleaf weeds in rice, exclusively.

EPTC (EPTAM®)

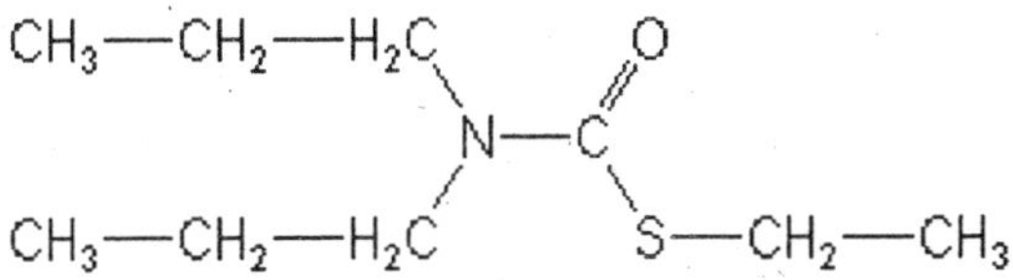

S-ethyl dipropylcarbamothioate

Molinate is effective for barnyard grass control in rice. Cycloate is used mostly in sugar beets, table beets, and spinach to control annual and perennial grasses and broadleaf weeds. Vernolate is effective for control of most grass weeds and some broadleafs in soybeans and peanuts. Butylate is incorporated preplant to control grass weeds in corn.

Two older thiocarbamates not illustrated, are diallate (Avadex®) and triallate (Far-Go®). Diallate (discontinued) is for wild oats and other grass weeds in a wide variety of crops. Triallate controls only wild oats in small grains.

Triazines

The *triazines*, which are six-member rings containing three nitrogens (the prefix *tri-* means "three") and azine (a nitrogen-containing ring) make up the heterocyclic nitrogens.

Triazines are strong inhibitors of photosynthetic electron transport. Their selectivity depends on the ability of tolerant plants to degrade or metabolize the parent compound (the susceptible plants do not). Triazines are applied to the soil primarily for their postemergence activity. There are many triazines on the market today, a few of which are illustrated.

Atrazine was estimated by EPA to have been the most heavily used pesticide in the U.S. in 1999, used at the annual rate of about 80 million pounds, primarily on corn. (Donaldson et. Al., 2002). In 2003 EPA concluded a key milestone in a special review of atrazine use that had gone on for many years. Though many uses of the product will continue a number of restrictions in use rates and pattern will apply in order to reduce the probability of atrazine residues moving to ground or surface water.

ATRAZINE (AATREX®)

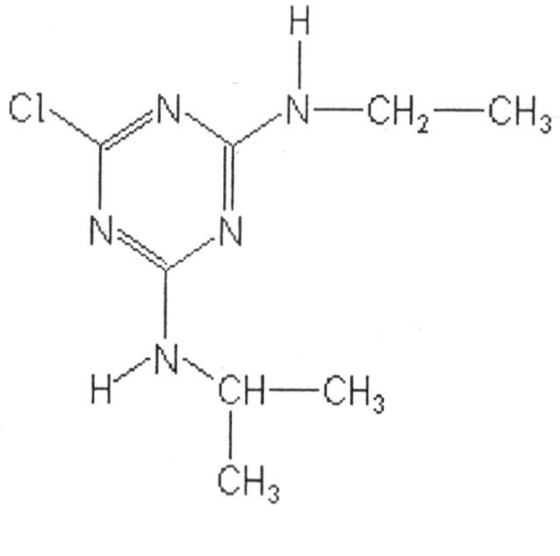

*6-chloro-*N*-ethyl-*N′*-(1-methylethyl)-1,3,5-triazine-2,4-diamine*

In a 1996 agreement with EPA uses for cyanazine were phased out by the manufacturer. Other important triazines not illustrated, are prometryn (Caparol®), ametryn (Evik®), and terbutryne (Igran®).

The triazines are used selectively in certain crops and nonselectively in others. They are used in greatest quantity in corn production and nonselectively on industrial sites.

Triazinones

The triazinones consist of two herbicides, metribuzin and hexazinone. Metribuzin is used on soybeans, wheat, sugarcane, and a few vegetables. Hexazinone is used in the control of annual, biennial, perennial weeds and woody plants on noncropland. It is sometimes applied as pellets or gridballs to control woody plants that fall to the ground through the dense foliage of conifer plantings e.g., Christmas trees.

METRIBUZIN (SENCOR®)

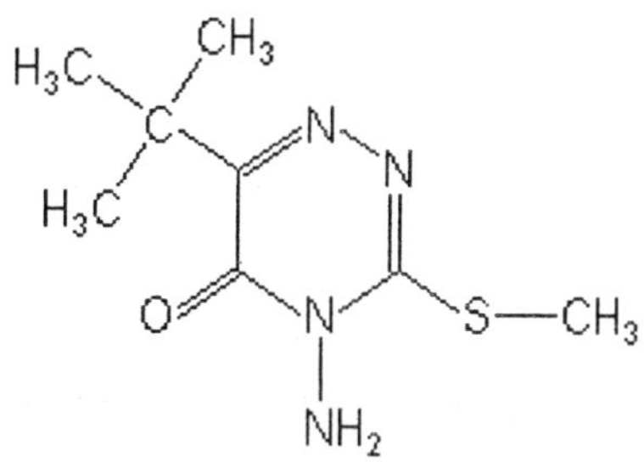

4-amino-6-(1,1-dimethylethyl)-3-(methylthio)-1,2,4-triazin-5(4H)-one

Triazoles

The oldest and only member of this group is amitrol, which became embroiled in the historic "cranberry incident" of 1959. This culminated in the addition of the Food Additive Amendment of 1958 (also called the Delaney clause) to the Federal Food, drug, and Cosmetic Act. In essence, this amendment stated that no residue of any carcinogen (cancer-producing agent) would be permitted in a food crop.

Uracils

Below is illustrated the uracil nucleus, and on it are substituted several chemical groups or moities, to form the herbicides bromacil, terbacil, and a third sold only in Europe, lenacil. Bromacil is a pre- and postemergence herbicide, for noncrop usage, pineapples, and citrus. It controls a very wide spectrum of grass and broadleaf weeds when applied early in the growing season, and must be carried into the root zone by moisture to be absorbed by the roots. Terbacil controls most annual grasses and broadleaf weeds when applied as a selective, preemergence herbicide. As with bromacil, moisture is required to move the material into the weed germination zone by sprinkler irrigation or rainfall. Terbacil has an extended soil residual life and non-tolerant crops should not be planted in treated fields for two years, nor should terbacil be used in sandy soils or those with less than 1% organic matter.

BROMACIL (HYVAR®)

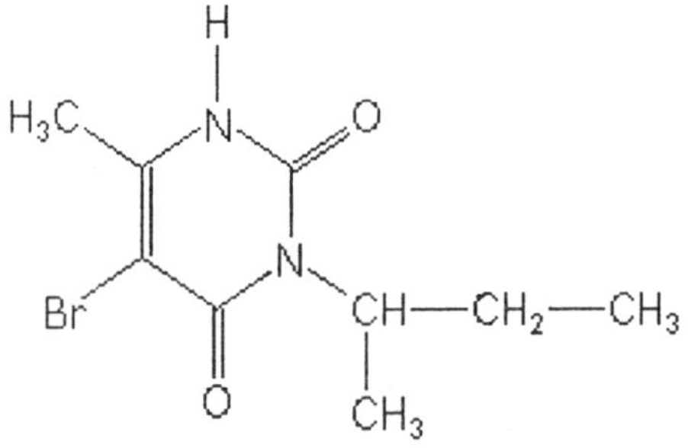

5-bromo-6-methyl-3-(1-methylpropyl)-2,4(1H,3H)-pyrimidinedione

The uracils inhibit photosynthesis by blocking the Hill reaction, as do the ureas and triazines. They are usually applied to the soil and move in the transpiration stream.

New Novel Herbicide Classes

A number of new products representing structures that do not easily fit into those classes covered above have appeared and are discussed below by class.

Benzoylcyclohexanediones- The three members of this class are mesotrione (Callisto®), sulcotrione (Mikado®) and benzofenap (Yukawide®). Mesotrione is the only of this group used in the U.S. and it was first registered in 2001 for use in corn to control broadleaf weeds and crabgrass. Sulcotrione is used in maize and sugarcane and benzofenap in rice. All of these materials are inhibitors of 4-hydroxyphenyl-pyruvatedioxygenase (4-HPPD).

MESOTRIONE (CALLISTO®)

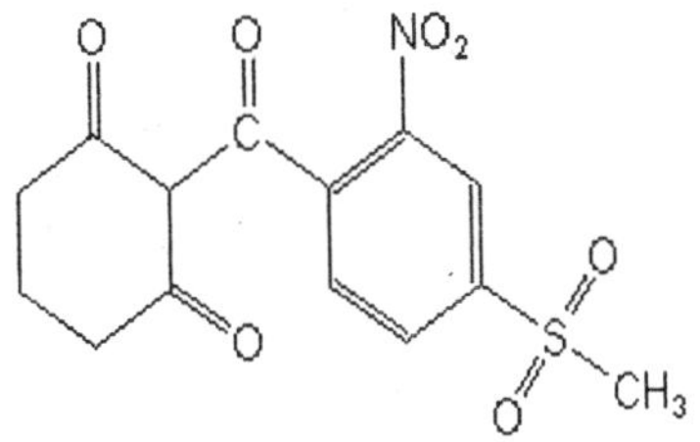

2-[4-(methylsulfonyl)-2-nitrobenzoyl]-1,3-cyclohexanedione

N-phenyl-phthalimides-Cinidon-ethyl (Lotus®, Bingo®) is used to control broadleaf weeds in small grains outside the U.S. This product acts by inhibiting protoporphyrinogen (PPO).

CINIDON-ETHYL (LOTUS®)

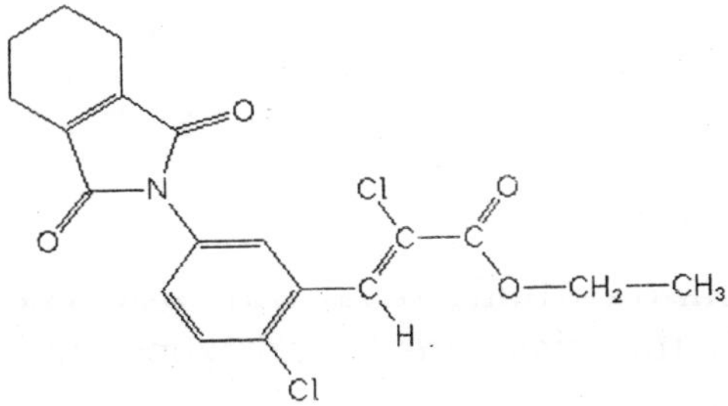

ethyl (2Z)-2-chloro-3-[2-chloro-5-(1,3,4,5,6,7-hexahydro-1,3-dioxo-2H-isoindol-2-yl)phenyl]-2-propenoate

Oxazolidinediones- Pentoxazone (Wechser®) is the only member of this class, at present. It is used on rice and at present only in S. Korea. It is an inhibitor of PPO.

PENTOXAZONE (WECHSER®)

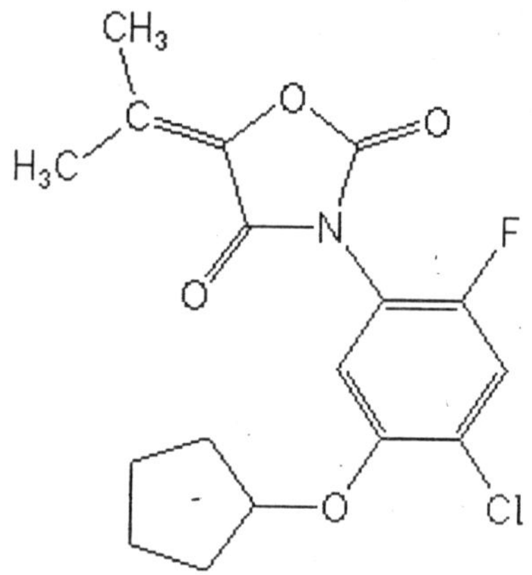

3-[4-chloro-5-(cyclopentyloxy)-2-fluorophenyl]-5-(1-methylethylidene)-2,4-oxazolidinedione

Phenyl-pyrazoles-The only member of this class is pyraflufen-ethyl (Ecopart®) and is now used outside the U.S. but has pending uses for broadleaf control in cotton, sugarcane, wheat, corn, potato and soybean. It is a PPO inhibitor.

Pyrimidindiones-Similar to the triazolinones class butafenacil (Rebin®, Inspire®) is the only member. It is used for weed control in grapes, nut crops, pome and stone fruits and also as a cotton defoliant. Its action is as a PPO inhibitor. Butafenacil was registered by EPA in 2003.

BUTAFENACIL (INSPIRE®)

Sulfonyamino-carbonyl-triazolinones- The two new entrants in this class are flucarbazone (Everest®) and propoxycarbazone (Attribute®, Olympus®) each of which exists as sodium salts. Each of these products is used in small grains to control grassy annual and some broadleaf weeds. Flucarbazone is registered on wheat in Canada and the U.S. Both products are inhibitors of acetolactate synthetase like the sulfonylureas.

FLUCARBAZONE (EVEREST®)

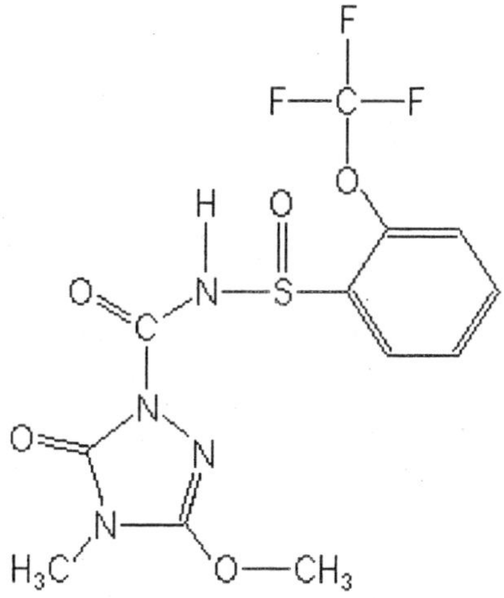

*4,5-dihydro-3-methoxy-4-methyl-5-oxo-*N*-[[2-(trifluoromethoxy)phenyl]sulfonyl]-1*H*-1,2,4-triazole-1-carboxamide*

Tetrazolinones-fentrazamide (Lecs®) is structurally reminiscent of the amide and chloroacetamid herbicides and is the only member of this class. It is a low toxicity material which is used to control barnyardgrass and annual sedges in rice in Japan and Korea. Like many of the amides and chloroacetamides it acts by inhibiting very long chain fatty acid synthesis,

FENTRAZAMIDE (LECS®)

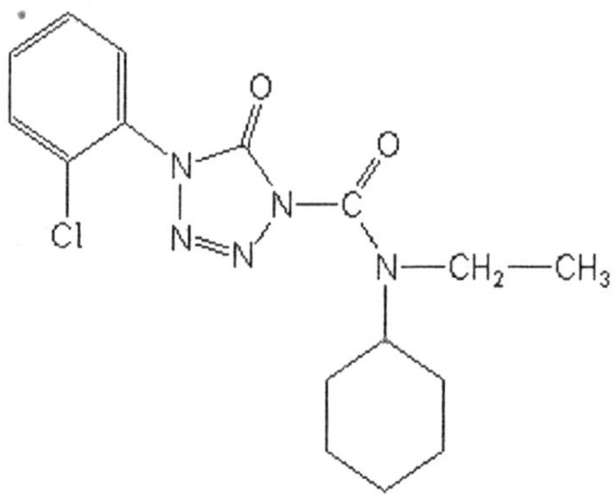

*4-(2-chlorophenyl)-*N*-cyclohexyl-*N*-ethyl-4,5-dihydro-5-oxo-1*H*-tetrazole-1-carboxamide*

Triazolinones- Amicarbazone (Dinamic®) is used to control broadleaf weeds in corn and soybeans and perhaps sugarcane. It is an inhibitor of photosystem II and is slated for a maize registration in the U.S. as early as late 2004.

AMICARBAZONE (DINAMIC®)

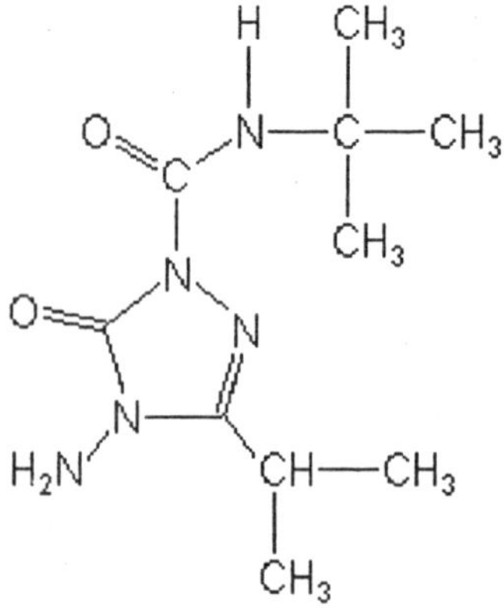

*4-amino-*N*-(1,1-dimethylethyl)-4,5-dihydro-3-(1-methylethyl)-5-oxo-1*H*-1,2,4-triazole-1-carboxamide*

Unclassified or Miscellaneous

Three herbicides have recently appeared that do not fit any particular established structural class. There are flufenpyr-ethyl (Axiom®), oxoziclomefone (Homerun®, Patful®) and benzobicyclon (Showace®). Flufenpyr-ethyl was registered by EPA in 2003 for corn, soybeans and sugarcane. It is a PPO inhibitor. Oxoziclomefone and benzobicyclon are both for use on rice in Japan. Their modes of action are not yet known.

Soaps

The principal value of soaps is in their capacity to reduce the surface tension of water, thus causing the guard cells surrounding the stomata to collapse. These in turn close or clog the stomata, the oxygen:carbon dioxide-exchange organs of the plants, causing death to the surrounding cells.

Soaps are the alkali salts of fatty acids. Both soft soaps (potassium salts) and the hard soaps (sodium salts) are water soluble, but only the soft soaps are herbicidal. Overall, the most efficacious fatty acid salts are those near the carbon chain length of lauric acid (C12), which include caproic (C10), myristic (C14), palmitic (C16) and stearic acids (C18).

In the late-1980s the first soft soap herbicides were registered. The two major manufacturers are the Ringer Corporation, which produces Safer® Moss & Algae Killer, and the Mycogen Corporation, which produces DeMoss® Moss/Algicide, and Sharpshooter® Herbicide, all aimed primarily toward the home yard and turf market.

Petroleum Oils

The earliest organic herbicides were the petroleum oils, which are a complex mixture of long-chain hydrocarbons. They include alkanes, alkenes, and often alicyclics and aromatics, produced by distillation and refinement of crude oils. The petroleum oils exert their lethal effect by penetrating and disrupting plasma membranes, which is enhanced by temperature and direct sunlight. Because of EPA's and many local and state governments' regulations to halt the release of hydrocarbons into the atmosphere, petroleum oils are no longer prominently used as herbicides though they are used as adjuvants in herbicide formulations.

Herbicide Safeners

It has long been known that mixing certain substances with herbicides before application will impart a protective or safening effect on selected crops without substantively affecting the weed control achieved. Few such agents were widely used before the last decade, however. Recently, the search for such substances has increased as producers have sought incremental performance and market advantages for their products.

Herbicide safeners are a diverse collection of compounds and many appear to impart their effects by enhancing the metabolism of the herbicide by the monocotyledonous crops on which the herbicide is used.

A few prominent herbicide safeners include: Cloquintocet-mexyl used with clodinafop-propargyl in Discover® Herbicide to safen the active ingredient on wheat. This safener and the herbicide with which it is used were first registered in the U.S. in 2000. Benoxacor, registered in 1997, increased the tolerance of s-metolachlor (and metolachlor) to corn. Dichlormid increases corn tolerance to thiocarbamate herbicides. Fenchlorazole-ethyl synergizes the activity of fenoxyprop-ethyl in susceptible species including wheat where the main action is to reduce phytotoxicity and growth retardation. Similarly, mefenpyr-diethyl safens fenoxyprop-P-ethyl in small grains. Fluxofenim and oxabetrinil both safen sorghum from metolachlor injury when applied as a seed treatment, particularly when metolachlor is used in combination with a triazine herbicide. Furilazole is an herbicide safener designed for wide range of herbicides on cereals, corn and rice. Finally, isoxadifen-ethyl is a safener for corn which has only very recently appeared.

Biopesticides

In addition to herbicidal agents of synthetic chemical origin, there are also a few so called "biopesticides", which are natural or semi-natural agents or organisms that can exact a degree of control over specific pests. A number of microorganisms (bacteria, viruses and fungi) and a few insects (e.g., parasitic wasps and the painted lady butterfly) have been evaluated as bioherbicides. In general, the only category that has yielded commercial examples, and these are few, are fungi (mycoherbicides).

The first to be registered was *Phytophthora palmivora* (Devine®) which induced selective disease on milkweed vine. Another commercial example is Collego®, living spores of the fungus *Collectotrichum gloeosporioides F. sp. aeschynomene*. It is applied to control coffee weed in rice or northern joint-vetch in soybeans. This fungus requires several weeks to achieve control, and to work at all the crop must be misted or the weather very humid.

Alternaria cassiae, though dropped as a commercial prospect, is also a myco-herbicide. It showed promise in controlling sicklepod in several crops.

Two other developmental myco-herbicides are: *Puccinia canaliculata* and *Xanthomonas campestis* for control of yellow nutsedge and annual bluegrass, respectively. Canada is exploring a new mycoherbicide, *Phoma macrostoma,* for use in turf.

TRANSGENIC PESTICIDES (HERBICIDE-TOLERANT CROPS)

Transgenic organisms are genetically altered by artificial introduction of DNA from another organism. The artificial gene sequence is referred to as a *transgene*. Plants with such transgenes are also referred to as being *genetically modified* (GM). Plants that emulate insecticides are those altered to induce *insect-resistance* (also called *plant pesticides* or *plant incorporated protectants)*. The other category of transgenic pesticides that we will touch on below is *herbicide-tolerant crops*.

Herbicide-tolerant crops are genetically engineered to sustain the action of selected herbicides applied directly over the young plants during standard weed control treatments. The significant commercial introduction of herbicide-tolerant crops began in the mid 1990's and prominent ones include corn, cotton and soybeans (among many others) designed to withstand the action of non-selective herbicides like glyphosate or glufosinate. Such tolerant crops allow broad spectrum weed control and high crop safety. Notwithstanding, there have been a number of crops engineered to induce tolerance to several other herbicides. Examples include bromoxynil, several imadazolinone and sulfonylurea herbicides, among others.

REFERENCES

Carpenter M.J, Ware G.W., *Defending Pesticides in Litigation*, 14th Ed., West Group, St. Paul, MN, 2004.

Donaldson D, Kiely T, Grube A., *Pesticides Industry Sales and Usage*, 1998 and 1999 Market Estimates. US EPA, Washington, DC, 2002.

Herbicide Classification of the Weed Science Society of America, *Weed Technology*, 1997.

10

Biological Control of Insect Pests

In recent decades, elevated awareness of the impacts of pesticide use on the environment and human health have resulted in efforts to reduce reliance on chemical controls. Many countries have instituted more stringent regulation of pesticide manufacture, registration and use, thereby increasing the cost, and decreasing the availability of these tools. In many cases, the pests themselves have indicated the need for change, with pesticide resistance now a common reality in many weeds, insects and diseases.

The use of natural enemies to reduce the impacts of pests has a long history. The ancient Chinese, observing that ants were effective predators of many citrus pests, augmented their populations by taking their nests from surrounding habitats and placing them into their orchards. Today's insectaries and air-freight delivery of natural enemies across the country or around the world are simply modern adaptations of these original ideas.

APPROACHES TO THE BIOLOGICAL CONTROL

There are three general approaches to biological control; importation, augmentation and conservation of natural enemies. Each of these techniques can be used either alone or in combination in a biological control program.

Importation

Importation of natural enemies, sometimes referred to as classical biological control, is used when a pest of exotic origin is the target of the biocontrol program. Pests are constantly being imported into countries where they are not native, either accidentally, or in some cases, intentionally. Many of these introductions do not result in establishment or if they do, the organism may not become pests. However, it is not uncommon for some of these introduced organisms to become pests, due to a lack of natural enemies to suppress their populations. In these cases, importation of natural enemies can be highly effective.

Once the country of origin of the pest is determined, exploration in the native region can be conducted to search for promising natural enemies. If such enemies are identified, they may be evaluated for potential impact on the pest organism in the native country or alternatively imported into the new country for further study. Natural enemies are imported into the US only under permit by the US Department of Agriculture. They must first be placed in quarantine for one or more generations to be sure that no undesirable species are accidentally imported (diseases, hyperparasitoids etc.). Additional permits are required for interstate shipment and field release.

Biological control of the alfalfa weevil, *Hypera postica* (Gyllenhall) is a example of a successful program using importation of natural enemies. The alfalfa weevil, a native of Europe, was originally detected in the US in Utah in 1904. A second introduction was detected on the East coast in 1951. By 1970, the weevil had spread to all 48 contiguous states and become a serious pest of alfalfa. Some importation's of natural enemies began as early as 1911, however, a major program aimed at biological control of the weevil was initiated in 1957. In this program, USDA ARS personnel conducted foreign exploration in Europe resulting in the eventual importation of 12 parasitoid species. Six of these species became established and are credited with contributing to the reduction in the weevil's pest status in the eastern US.

Augmentation

Augmentation is the direct manipulation of natural enemies to increase their effectiveness. This can be accomplished by one, or both, of two general methods: mass production and periodic colonisation; or genetic enhancement of natural enemies. The most commonly used of these approaches is the first, in which natural enemies are produced in insectaries, then released either inoculatively or inundatively. For example, in areas where a particular natural enemy cannot overwinter, an inoculative release each spring may allow the population to establish and adequately control a pest. Inundative releases involve the release of large numbers of a natural enemy such that their population completely overwhelms that of the pest. Augmentation is used where populations of a natural enemy are not present or cannot respond quickly enough to the pest population. Therefore, augmentation usually does not provide permanent suppression of pests, as may occur with importation or conservation methods.

An example of the inoculative release method is the use of the parasitoid wasp, *Encarsia formosa* Gahan, to suppress populations of the greenhouse whitefly, *Trialeurodes vaporariorum* (Westwood), (Hussey & Scopes 1985, Parrella 1990). The greenhouse whitefly is a ubiquitous pest of vegetable and floriculture crops that is notoriously difficult to manage, even with pesticides. Releases of relatively low densities (typically 0.25 to 2 per plant, depending on the crop) of *E. formosa* immediately after the first whiteflies are detected on a greenhouse crop can effectively prevent populations from developing to damaging levels.

However, releases should be made within the context of an integrated crop management program that takes into account the low tolerance of the parasitoids to pesticides.

Conservation

In any biological control effort, conservation of natural enemies is a critical component. This involves identifying the factor(s) which may limit the effectiveness of a particular natural enemy and modifying them to increase the effectiveness of the beneficial species. In general, conservation of natural enemies involves either, reducing factors which interfere with natural enemies or providing resources that natural enemies need in their environment.

Many factors can interfere with the effectiveness of a natural enemy. Pesticide applications may directly kill natural enemies or have indirect effects through reduction in the numbers or availability of hosts. Various cultural practices such as tillage or burning of crop debris can kill natural enemies or make the crop habitat unsuitable. In orchards, repeated tillage may create dust deposits on leaves, killing small predators and parasites and causing increases in certain insect and mite pests.

In one study, periodic washing of citrus tree foliage resulted in increased biological control of California red scale, *Aonidiella aurantii* (Maskell) due to increased parasitoid efficiency. Finally, host plant effects such as chemical defenses which are harmful to natural enemies but to which the pest is adapted, can reduce the effectiveness of biological control. Some pests are able to sequester toxic components of their host plant and use them as defense against their own enemies. In other cases, physical characteristics of the host plant such as leaf hairiness, may reduce the ability of the natural enemy to find and attack hosts.

Ensuring that the ecological requirements of the natural enemy are met in the cropping environment is the other major means of conserving natural enemies. To be effective, natural enemies may need access to; alternate hosts, adult food resources, overwintering habitats, constant food supply, and appropriate microclimates. In a classic example, Doutt & Nakata determined that *Anagrus epos* Girault, principal parasitoid of the grape leafhopper, *Erythroneura elegantula* Osborne in California grape vineyards required an alternate host for overwintering. This host, another leafhopper, only overwintered on blackberry foliage in riparian areas, often quite distant from the vineyards. Vineyards close to natural blackberry stands experienced earlier colonisation by the parasitoid in the spring and better biological control. Wilson et al. found that French prune trees which harbor another overwintering host, could be planted upwind of vineyards and effectively conserve *Anagrus epos.*

APPLICATIONS OF BIOLOGICAL CONTROL

Biological control is an exciting science because it constantly incorporates new knowledge and techniques. In this section we will illustrate several ways in which time honored approaches to biological control are being adapted to meet today's pest management challenges.

Modern Approaches in Augmentation of Natural Enemies

Because most augmentation involves mass-production and periodic colonisation of natural enemies, this type of biological control has lent itself to commercial development. There are hundreds of biological control products available commercially for dozens of pest invertebrates, vertebrates, weeds, and plant pathogens.

The practice of augmentation differs from importation and conservation in that making permanent changes in a agroecosystem to improve biological control is not the primary goal. Rather, augmentation generally seeks to adapt natural enemies to fit into existing production systems. For example, cultures of the predatory mite, *Metaseiulus occidentalis* were laboratory-selected for resistance to pesticides commonly used in an integrated mite management program in California almond orchards. This program has saved growers $24 to $44 per acre per year in reduced pesticide use and yield loss. Genetic improvement of several predators and parasitoids has been accomplished with traditional selection methods, and appears possible with recombinant DNA technology.

An excellent example of an augmentative practice than has been successfully adapted to a wide variety of agricultural systems is the inundative release of *Trichogramma* wasps. These minute endoparasitoids of insect eggs are released in crops or forests in large numbers timed to the presence of pest eggs. *Trichogramma* are the most widely augmented species of natural enemy, having been mass-produced and field released for almost 70 years in biological control efforts. Worldwide, over 32 million ha of agricultural crops and forests are treated annually with *Trichogramma* spp. in 19 countries, mostly in China and republics of the former Soviet Union.

In China, agricultural production and pest management systems capitalise on low labor costs, and generally follow highly innovative yet technologically simple processes. For example, *Trichogramma* spp. that are inundatively released to suppress sugarcane borer, *Chilo* spp., populations in sugarcane are protected from rain and predators inside emergence packets. Insectary-reared parasitised eggs are wrapped in sections of leaves which are then slipped by hand over blades of sugarcane. Most *Trichogramma* production in China takes place in facilities producing material for a localised area. These facilities range from open air insectaries to mechanised facilities that are leading the world in development of artificial host eggs.

One of the barriers to wider implementation of biological control in western agriculture has been socioeconomics. In current large-scale production agricultural systems, a premium is placed on efficiency and economy of scale. Entire support industries have developed around the application of agrichemicals, including application equipment manufacturing, distribution and sales, as well as application services. In order for biological control products to not be at odds with these industries, and to compete strongly with pesticides, they should have many of the same characteristics. Ideally, they should be as effective as pesticides, have residual activity, be easy to use, and they should have the capacity to be applied quickly on a large scale with conventional application equipment.

In Western Europe, almost two decades of intensive research resulted in the commercial marketing of three products utilising the European native, *Trichogramma brassicae* Bezdenko, to suppress the European corn borer, *Ostrina nubilalis* Hübner, in corn fields. These products are annually applied to approximately 7,000 ha in each of Switzerland and Germany, 150 ha in Austria, and 15,000 ha in France. All three products are based on manufactured plastic or paper packets designed to provide protection for the wasps against weather extremes and predation until emergence in the field.

As in the Chinese example above, European *Trichogramma* products are for the most part applied to crop fields by hand. One exception is the product called, Trichocaps which can be broadcast either by hand or by aircraft using conventional application equipment. Trichocaps packets are actually hollow walnut-shaped cardboard capsules that each contain approximately 500 parasitised Mediterranean flour moth, *Ephestia kuehniella* Zwolfer, eggs. Developing *Trichogramma* inside capsules are induced into an overwintering (diapause) state in the insectary, then stored in refrigerated conditions for up to nine months without loss of quality. This system allows for production of product during winter months, then distribution to growers when needed in the summer.

Once removed from cold storage, *Trichogramma* inside the capsules will begin development and begin emergence approximately 100 Celsius degree days later. This 'reactivation' process can be manipulated so that capsules containing *Trichogramma* at different developmental stages can be applied to fields at the same time, extending the emergence period of parasitoids and increasing the 'residual' activity of a single application to approximately one week. Planning and preparation of the product for application is done by the company so that growers are only responsible for applying the product to crop fields.

Cooperative research over the last 5 years has resulted in successful commercial-scale pilot testing of this method in North America on seed corn and field corn production systems. This strategy now has the potential for immediate commercial implementation in North America.

Landscape Ecology and the Conservation of Natural Enemies

The study of disturbance and its effects on community dynamics and the emergence of the discipline of landscape ecology are impacting the way we think about the conservation of natural enemies. Over the past 15 years, ecologists have come to recognise the central role that disturbance plays in the structuring of ecological communities. While the most highly disturbed terrestrial ecosystems may have one disturbance event every several years (e.g. fire in grasslands), many agricultural ecosystems experience multiple events per growing season (plowing, planting, nutrient and pesticide applications, cultivation and harvest). From an ecological point of view, the outcomes are predictable. Highly disturbed systems exhibit decreased species diversity and shortened food chains, resulting in the few well adapted species (i.e. pests) having few natural enemies to suppress their populations. This requires that additional disturbance events be initiated (i.e. pesticide applications) which, while controlling the initial negative symptom, may precipitate its reoccurrence.

Current systems of crop production also shape the physical structure of our agricultural landscapes . With increased reliance on mechanisation and pesticides, diversity in farmlands has rapidly disappeared and the impacts on natural enemies are only now beginning to be understood. In general, increased habitat fragmentation, isolation and decreased landscape structural complexity tend to destabilise the biotic interactions which serve to regulate natural ecosystems.

The goal of an ecological approach to conservation biological control is to modify the intensity and frequency of disturbance to the point where natural enemies can function effectively. This will need to occur at field, farm and larger landscape-levels. Within fields, modification of tillage intensity and frequency (reduced tillage or no-tillage) can leave more plant residue on the soil surface and have a positive impact on predators (ground beetles and spiders). Intercropping can also modify the microclimate of crop fields making them more favorable for parasitoids.

At the farm level, the presence and distribution of non-crop habitats can frequently be critical to natural enemy survival. *Eriborus terebrans* (Gravenhorst) is a wasp which parasitises European corn borer larvae. Female *Eriborus* require moderate temperatures (<90° F) and a source of sugar (nectar of flowering plants or aphid honeydew). Neither of these conditions is met in a conventionally managed corn field. Therefore, wasps seek more sheltered locations in wooded fencerows and woodlots where they find reduced temperatures, higher relative humidity and abundant sources of adult food. European corn borer larvae in corn field edges near these types of habitats are parasitised at two to three times the rate of those in field interiors (up to 40%). Current research is examining the potential of modifying corn production systems by creation of natural enemy resource habitats to provide critical resources and increase natural control of European corn borers. Intercrops, strip crops, as well as modification of grass waterways, shelterbelts, buffer and riparian zones are promising techniques.

Finally, at the landscape-level, the physical structure of agricultural production systems can also influence pest and natural enemy diversity and abundance. In a study contrasting simple versus mosaic landscapes, Ryszkowski et al. concluded that natural enemies are more dependent on refuge habitats than are pests and the greater abundance of these refuges in the mosaic landscapes resulted in their higher diversity, abundance and ability to respond to prey numbers. Marino and Landis examined parasitism of true armyworm, *Pseudaletia unipuncta,* in structurally-complex versus simple agricultural landscapes. Overall parasitism in the complex sites was more than three times higher than in the simple sites (13.1% versus 3.4%). Differences were largely attributable to one wasp species, the braconid, *Meterous communis* which was far more abundant in complex sites. They hypothesised that abundance and proximity of preferred habitats for alternate hosts of *M. communis* may account for the observed differences.

In the past, conservation was typically attempted one species at a time, concentrating on meeting the needs of what was deemed the most important natural enemy in a particular system. While this will continue to be an enormously useful approach, it now seems possible that basic ecological theory could inform the design and management of landscapes to conserve and enhance the effectiveness of entire communities of natural enemies.

PREY SPECIALISATION IN INSECT PREDATORS

Prey Specificity and Pest Management

Contrary to common assumptions, many predacious

insects are not generalist feeders; some are conspicuously selective in their diet. Dietary breadth represents a key component in the ecology, behavior, and evolutionary diversification of predators. It is also a central issue in the manipulation and use of predators in IPM. For example, prey specificity impinges directly on the use of predators in all types of biological control — importation, as well as augmentation and manipulation. Biological control workers generally favor using host-specific natural enemies, and in choosing natural enemies, they often consider a high level of host specificity as a prime attribute. Indeed, the first modern success in biological control involved a highly specific predator, the vedalia beetle, *Rodolia cardinalis*. However, prey specificity is not always necessary in pest management; many predatory species that are currently mass-produced and released in field crops and orchards, have relatively broad prey ranges; examples include the anthocorids *Orius insidiosus* and *O. tristicolor* and the green lacewings *Chrysoperla carnea* and *C. rufilabris*. The desire for dietary specificity presents a dual challenge for classical biological control programs involving predators. On the one hand, the interaction between the predator and target pest should be predictable; on the other hand, the nutritional requirements for an introduced predator may be complex and may include factors that are not readily ascertained, e.g. acceptance or preference for novel prey in the new habitat, obtaining nutrients from symbiotic microorganisms, utilisation of supplementary food, such as plant fluids, fungi, pollen, or nectar. A predator's use of these dietary resources has important implications for the outcome and stability of predator-prey dynamics. Given the broad range of variability in the diet breadth of predacious insects, biological control efforts would benefit greatly from detailed investigations of predator-prey interactions; such studies will become a necessity in the near future.

Prey Specificity and Environmental Conservation

The release of insect predators with relatively broad host ranges raises significant questions about negative effects on non-targeted organisms. It is possible that introduced predators may move out of the target crop area and prey on species other than pests (possibly rare or threatened species). To our knowledge, adverse effects from the introduction of insect predators have been examined for only one species. The dispersal of the polyphagous aphid-feeding coccinellid, *Coccinella septempunctata,* has been associated with reduced abundance of two indigenous coccinellid species in South Dakota and possible adverse effects on the biological control of the alfalfa weevil in Utah. Thus, assessment of diet breadth should be an integral part of the pre-release evaluation of entomophagous species, especially when rare or threatened species may be at risk. In this regard, many aspects of host specificity in biological control programs were recently reviewed.

INSECT PREDATORS

Typically, insect predators are characterised by a set of attributes that distinguish them from parasitoids, the other major group of entomophagous insects. They are large relative to their prey and require more than one prey individual to complete development; they have free-living predatory immature stages; and many species of insect predators are predacious as both immatures and adults. For most predacious insects, prey consists of other insects, but some may consume animals in other classes. Except for predatory Hymenoptera that provision their nests with prey, predacious insects consume their prey immediately after attack. Predators occur in approximately 20 insect orders; lists vary slightly, depending on the definition of scavenging and fortuitous predation. The only orders of insects not known to contain predatory species are the

Isoptera, Phasmatodea, Phthiraptera, Strepsiptera, and Siphonaptera.

Feeding Habits

Predacious insects exhibit great variation in their dietary range. Some, e.g, the vedalia beetle, Rodolia cardinalis, and the green lacewing Chrysopa slossonae, are highly specific and feed on only one species of prey. Others, e.g. the aphid-feeding coccinellids *Hippodamia convergens* and Adalia bipunctata, are stenophagous or oligophagous and restrict feeding to a range of related taxa. Finally, other predators, e.g. the bug *Podisus maculiventris,* and the lady beetle *Coleomegilla maculata,* are polyphagous (general feeders) and consume a wide variety of prey and non-prey items, e.g. plant fluid, pollen. It is noteworthy that many species with wide prey ranges contain biotypes and populations that differ in their responses to prey; examples include web-building and jumping spiders and the green lacewing Chrysopa quadripuncata. In certain cases, symbiotic microorganisms may supply nutrients for development or reproduction. For example, the common green lacewing Chrysoperla carnea harbors yeasts (Torulopsis) in an enlarged esophageal diverticulum; these yeasts can provide essential amino acids that are lacking in the lacewing's diet.

Predators can be classified according to the life stage of prey they attack (e.g., egg predators), their foraging strategy (e.g., active searchers, ambush or filter feeders, web or bolas builders). Although these types of categories may be generally informative about the type of prey taken, they have no predictive value for the prey specificity of individual species. In this regard, it is of both practical and theoretical significance that predator phylogeny can be an important clue to prey breadth and prey preference. For example, among the Coccinellidae, the Chilocorninae prey on Homopteran scale insects, most species of Coccinellinae are aphid predators, and the Stethorinae has specialised on phytophagous mite species. However, it is critical to note that sister-species may differ in the range of prey that they take within their preferred type of prey; e.g., sister-species may include a generalist and a specialist aphid-feeder. Therefore, care must be exercised in using phylogenetic relatedness to generalise about the prey breadth of specific taxa.

Methods for Determining Prey Range

A major challenge in IPM is to assess the action of biological control agents. Unlike parasitoids that leave the exoskeleton of their host, predators often leave no clues after feeding. Observation and gut analysis have been used for a long time to determine what predators eat. For chewing-type predators, analysis of feces can also indicate what prey were in the diet. Considerable advancement has taken place recently in the development of serological techniques for identifying which predators feed on target pests in the field; ELISA and immunodot assays are relatively rapid, inexpensive and easy to interpret. Although quantifying the prey consumed presents significant technical problems, the tests are useful for determining key predators within agricultural systems and for evaluating the efficacy of augmentative biological control agents.

ELEMENTS OF PREY SPECIFICITY

Predators that are capable of taking a very catholic diet (e.g., in the laboratory) often take a much narrower range of food. As in the case of herbivores and parasitoids, diet-breadth in predators results from the interaction of diverse physiological, behavioral and ecological factors: (a) the relative availability of specific types of prey, (b) the foraging behavior of predators, (c) the suitability of prey, and (d) the risk of predation or other mortality factors associated with obtaining prey. In many cases parental behavior, especially oviposition behavior, can play a crucial role in determining the prey that are available to predacious larvae.

Prey Availability

Prey abundance can determine whether a predator enters or remains in a habitat, as well as the type of prey and relative numbers of prey that the predator consumes. A predator is said to show a preference for a specific type of prey if the percentage of that type of prey in its diet is higher than the proportion of the prey available to it. Although some predators' prey preference is fixed (maintained irrespective of relative availability in the environment), other predators may switch to more common prey species. Seasonal availability of prey can also be a major determinant of prey specificity. Predators show several adaptations to low prey densities; these include changing trophic levels (e.g., intra-guild predation, cannibalism), movement to areas with higher prey densities, switching to new prey, and entering prey-mediated dormancy.

Predator Foraging Behavior

Habitat & prey location by insect predators

Adult predators may use a variety of cues, including volatile chemicals from both plants and insects, to find prey. The role of chemical cues in locating prey is relatively well understand for the green lacewing *Chrysoperla carnea*. A complex mixture of volatiles (an attractant and an arrestant) induce *C. carnea* adults to fly upwind, toward a food source, and having reached the food, to cease movement and to oviposit. One the volatile chemicals is a metabolite of oxidised tryptophan (most likely indole acetaldehyde) which occurs in aphid honeydew. In addition, for the adults to respond to the tryptophan in honeydew, they must perceive a volatile synomone from plants, caryophyllene, which is a relatively common volatile from cotton and alfalfa plants. Knowing the influence of these compounds on predator behavior forms a strong basis for manipulating these natural enemies in the field. Their application attracted large numbers of lacewings to cotton and potato fields. Several species of predacious beetles and bolas spiders also use chemical cues (plant terpenes, prey pheromones, or mixtures of terpenes and pheromones) to find their prey. Larval predators have evolved various means of locating and recognising their prey: phototactic and/or geotactic responses, vision, olfaction, sound or vibration detection, contact. For example, predacious mites use chemical cues to find their prey. Searching behavior may be altered by encountering prey; some larval predators switch from linear movement to area-intensive searching after they contact prey. A predator's previous experience (learning) can influence its searching behavior, as well as the type and proportion of prey taken.

Plant structure may also affect the pattern of predator movement. Species and biotypes of predators can differ radically in their habitat associations and in the microhabitats they choose. For example, certain ladybird beetles occur in plant canopies, whereas others inhabit areas closer to the soil. *Chrysoperla rufilabris* prefers moist habitats, whereas *C. carnea* is well adapted to dry situations. These differences in habitat association are reflected in the prey available to predators and they are a very important factor in the choice of species or biotypes to be used in specific biological control situations. Prey Acceptance.

After the predator discovers prey, it must pursue, subdue, and consume it. These three functions constitute the predator's handling time, which is a very important determinant of diet breadth. By definition, generalist predators attack, subdue and consume a wide range of the prey species they encounter. Several morphological and physiological factors can influence prey acceptance. One underlying physical factor is the size of prey. After contact with a potential prey, the characteristics of cuticle or the presence of waxes may affect the predator's responses to the prey. For example, prey acceptance by the vedalia beetle, which has a very restricted diet breadth, appears to be influenced by the waxy texture of its prey, the cottony cushion scale. Prey defenses (e.g., behavioral, morphological, chemical) may determine whether a predator accepts or rejects prey. Some prey (e.g. aphids) kick, run, drop, or fly away, or exude noxious chemicals when predators approach them. Sometimes, the chemicals induce the predators to vomit. Sih and Endler discuss the defensive interactions between predators and their prey.

Prey Suitability

If certain types of prey are not suitable (i.e., they have low nutritional quality for the predator), the predator may ultimately reject the prey, or it may continue feeding, but with detrimental effects. The negative effects include reduced rates of development, reproduction or survival. For example, several species of coccinellids and chrysopids exhibit slow development and reduced fecundity when they eat suboptimal aphid prey. In some cases, predators continue feeding when the prey contain toxins that result in the predator's death. Prey suitability may or may not be the same for immatures and adults. For example, successful development and reproduction in the green lacewing, *C. slossonae*, requires that both larvae and adults feed on the predator's specific prey (the woolly alder aphid). A diverse array of studies on the nutritional (dietary) requirements of predators have been conducted. Despite these studies, relatively few insect predators can be mass-reared successfully, and even fewer can be reared on artificial diets. The development of artificial diets for predators is a very significant area for the future development of biological control.

Natural Enemies Associated with Prey

Virtually all animals are vulnerable to natural enemies; active predators themselves often carry a very high risk of detection and attack by parasitoids or other predators. Thus, an important determinant of diet breadth is the range and abundance of natural enemies that are associated with the predator's food source. Predators' defenses include becoming immobile, living in protected places (ambush predators, such as pit-dwelling antlion larvae), cryptic coloration and polymorphisms (Hawaiian spiders and lacewing larvae), mimicry (mantispid-mimicing wasps), escape and threat behavior, and noxious chemical exudates. The evolution of a predator's defenses against natural enemies may result in trade-offs in terms of reduced time or efficiency in searching for, attacking, consuming, and metabolising prey.

EVOLUTION AND PREDICTABILITY OF PREY SPECIFICITY

Stability in predator-prey interactions constitutes a vital issue in IPM. To evaluate the long-term persistence of prey specificity, it is necessary to understand the genetic variation and evolution of predator-prey associations. Such an understanding requires analysis of the behavioral, physiological, phenological and other traits that underlie predators interactions with prey, as well as comparative experimental studies of predators within a phylogenetic context. In contrast to herbivores, very few such studies exist for predacious insects. Such studies would in;hance the further development of IPM.

Genetic Variation in Predator-Prey Interactions

Genetic variability in predator-prey relations have been examined for only a very few arthropod predators — the spider *Agelenopsis aptera* and the green lacewing *Chrysopa quadripunctata*. For the two species, differences in both foraging and defensive behavior are related to variation in prey availability and predation risks among habitats. In the case of the lacewing, the expression of the genetic variation in foraging and defensive behavior is enhanced by high levels of individual repeatability in the behavior, even after the predacious larvae moult twice. Genetic variation in the seasonal and habitat associations among *Chrysoperla carnea* populations has also been studied; the importance of the genetic variation to biological control and IPM is discussed by Tauber and Tauber and Luck et al.

Phylogenetic Relationships

An effective method for analysing the evolution and stability of food associations in insects is to conduct comparative studies among phylogenetically related taxa. Such an approach has revealed both patterns and processes in the evolution of insect herbivore associations with plants. In marked contrast, the link between phylogeny and prey associations remains relatively unexplored primarily because phylogenetic relationships within most groups of predatory insects are very poorly known. Nevertheless, some relatively recent comparative studies with syrphids and chrysopids have used a phylogenetic basis to investigate the evolution and stability of specific predator-prey associations. Such studies have strong basic and applied implications. Within the hoverflies (Diptera: Syrphidae), the subfamily Syrphinae is believed to be a monophyletic group, and all described species are predatory.

Using a phylogenetic framework, long-term population census data, and comparative data on morphological characteristics (e.g., body size, egg size and number), Gilbert and co-workers tested several hypotheses concerning specialisation of prey range. For aphid predators in the Syrphinae, the data do not support predictions that specialised predators lay few but large eggs, or that adults of specialised predators tend to be large. The results indicate that specialisation is a derived trait in the Syrphinae and that long-term stability of populations of specialist Syrphinae is similar to that of the generalists. Phylogenetically based investigations examined the evolution and stability of prey specialisation in the Chrysopidae. The family is considered monophyletic, but relationships within the tribe Chrysopini, which contains most species used in biological control, need further study. Comparative experimental data from a pair of sister-species (*C. quadripunctata*, a generalist, and *C. slossonae*, a derived specialist) demonstrate that a three-step process underlies the evolution of prey specialisation.

First, establishment of a predator population on a new food involves genetic variability or phenotypic plasticity in prey usage. Second, adaptation to the specific prey may involve behavioral, physiological, morphological and/or phenological traits. Unlike in the Syrphinae, the evolution of prey specialisation in *Chrysopa* is associated with increased egg size, reduced fecundity and increased body size; trade-offs among the traits are also important. Finally, the maintenance of prey specialisation requires the evolution of reproductive isolation between the ancestral and derived populations. In the case of *C. quadripunctata* and *C. slossonae*, this step included the evolution of seasonal asynchrony, perhaps specific prey requirements for mating, and a postmating, prezygotic barrier to fertilisation. Given the large differences and the high degree of reproductive isolation between the generalist and specialist sister-species, *C. slossonae*'s prey specialisation appears very stable.

BIOLOGICAL CONTROL OF CROP PESTS IN THE CARIBBEAN

Biocontrol of plant pests in the Caribbean was almost completely neglected with the advent of pesticides which appeared to be the solution to all pest problems. However, due to the development of resistance in some major pests, as well as to public awareness of hazards posed to human health and the environment searches are being made for seemingly old alternatives.

The tropical environment in the Caribbean facilitates year round growing season and supports continuous generations of both pests and beneficial organisms. Agriculture is characterised by small subsistence farms, multiple crops and pests, pesticide problems and other problems common to production and marketing.

Biocontrol is now being acknowledged as a area for further research and is a component of a USDA initiative in sustainable agriculture. The ARS, the Entomology Department Administrators in U.S. Universities and the Experiment Stations have proposed working groups to prepare national biocontrol research programs. This has evolved as an effective alternative to the pesticide dilemma.

Biocontrol is one of the most thoroughly studied methods against insects, nematodes, diseases and weeds. Scientists are now increasing efforts and emphasis on pest control methods not aimed at wholesale eradication, but rather at maintaining pest populations below economic thresholds.

This has already started in the Caribbean, where efforts of national and international institutions have begun to emphasize and implement biocontrol programs. These efforts are directed at identification of research priorities and possible areas for cooperation within the region. At present, there are several institutions conducting individual research with little or no regional coordination. Biocontrol efforts and findings remain little known even between neighboring countries. This is due in large measure to inadequate communication between research and development organisations in the region. There are limited means to disseminate and exchange findings. Lack of funds for research and for attending local or international meetings is by far the greatest constraint to the development of biocontrol in the region. A workshop should serve to identify research needs and to promote interchange of knowledge within the Caribbean community.

There is evidence of the effective biocontrol successes worldwide and in the Caribbean. Bennet and Alam et al. summarised the most important cases of successful biocontrol in the Caribbean. The recent introduction of the citrus blackfly *Aleurocanthus woglumi* in Puerto Rico and its outstanding effective control with the subsequent introduction of the parasitioids *Encarsia opulenta* and *Amitus hesperidum* are a pertinent example. Chemical control of this pest has been impossible in most places; only biocontrol has been effective.

In Puerto Rico, as in many other Caribbean countries, biocontrol has been used for decades. Table 1 presents information regarding the introduction of parasites and predators into Puerto Rico in the 1900's. Most introductions were made during the 1930's, up to 1940, just before the pesticide era. As often occurs, data on many of these introductions are not available since relevant studies were never conducted.

Some exotic natural enemies have been recently introduced for the control of certain insect pests, such as citrus blackfly, the lima bean pod-borer, and the sugarcane borer and others such as nematodes, diseases and weeds. Also, several species of entomophilic nematodes for the control of root weevils were introduced and tested.

Some recent biocontrol attempts have been successful in Caribbean countries, i.e. the sugarcane borer, the citrus blackfly, the diamondback moth (to a certain level). On-going projects are devoted to control the coffee berry borer, Hipothenemus hampei. Two potential parasitoid candidates have been identified: Prorops nasuta and

Cephalonomia stephanoderes. These parasitoids will be reared and increased for distribution in the areas where the pest has been identified: Colombia. Jamaica, Mexico, etc. Very positive results are expected from this initiative.

IMPORTANT CROPS AND INSECT PESTS IN THE CARIBBEAN

Among the most important crops grown in the Caribbean are coffee, sugarcane, bananas, plantains, root crops, tomato, peppers, cabbage, cucurbits, citrus, pineapple, avocado, mango, pigeonpea and beans. There are important insect pests attacking these crops, some of them with a high potential for biocontrol.

Sugarcane Borer

The sugarcane borer, Diatraea saccharalis, is under effective biocontrol in some countries, but it remains an important limiting factor in several others. Different natural agents have been successful on some countries, probably indicating the influence of varying habitats. This pest offers a good opportunity for classical biocontrol. Several natural enemies have been reported by Bennet and others. Currently, *Cotesia flavipes*, *Lixophaga diatraeae*, *Metagonistylum minense* and *Trichogramma* spp. are the most successful biocontrol agents. Research is needed to continue the improvement of the existing biocontrol agents.

Root Weevils

Historically, the sugarcane rootstalk weevil, *Diaprepes abbreviatus*, has been among the most difficult pest to control in the Caribbean. One contributing factor is the high reproductive potential (5,000 eggs/female), in a situation where very low egg survivorship is enough for it to be a pest. Several biological control attempts have failed, except for entomopathogenic nematodes. However, the search for other biocontrol agents continues Castro studied the ants species feeding on neonate larvae. She found seven feeding on the larvae. Of these, *Pheidole subarmata borinquensis* and *Pheidole fallax* demonstrated the greatest efficiency. In Cuba, Castineiras et al., reported *P. megacephala* controlling *Cosmopolites sordidus* and *Cylas formicarius*. Richman et al. reported the ants *Monomorium floricola* and *Crematogaster ashmeadi* preying on the egg masses of *D. abbreviatus* in Puerto Rico and Florida, respectively. Colón found 17 species of fungi associated with the larvae of D. abbreviatus. Pathogenicity tests demonstrated that half of then were able to kill the larvae. *Gliocadium* sp. and *Fusarium* sp. caused the highest mortality. Armstrong studied the efficiency of the egg parasitoid *Tetrastichus haitiensis* on *D. abbreviatus* eggs oviposited on sugarcane. He found a low efficiency of the parasitoid and concluded it was probably due to the difficulty of oviposition through the sugarcane leaves as compared to other hosts.

Several species of entomopathogenic nematodes have been tested for the control of the most important root weevils in Puerto Rico. Román & Figueroa reported effective control of the larvae of *D. abbreviatus* with *Steinernema feltiae (=Neoaplectana carpocapsae)*. Figueroa & Román reported tests with *S. glaseri, S. bibioni* and *Heterorhabditis heliothidis* against *D. abbreviatus*. González observed some control using H. bacteriophora and H. heliothidis. Figueroa obtained excellent results with *S. feltiae* and *S. biboni against C. sordidus*. A graduate student is currently evaluating *S. feltiae* and *H. heliothidis* against *C. formicarius*. Several local strains or species of entomopathogenic nematodes recently found in Puerto Rico are being evaluated. Natural enemies against other curculionid pests such as *Exophthalmus* spp. and *Pachnaeus* spp. are being sought in some Caribbean countries.

Coffee Berry Borer

The coffee berry borer, *H. hampei* is threatening coffee production in the Caribbean. Biocontrol research has produced encouraging finding. Besides the two parasitoids mentioned in the introduction, other potentially effective parasitoids are under study. An example is *Heterospilus coffeeicola,* which has been classified as effective but difficult to rear in the laboratory. Other potential candidates such as: Phymasticus coffea, *Aphanognus dictynnu* (apparently hyperparasite of *Prorops nasuta*) and some microbial agents such as *Beauveria bassiana, Metarhizium anisopliae* and *Paecilomyces* spp. These are under study at various research centers worldwide.

Coffee Leafminer

Another pest of coffee with a high potential for biological control is the coffee leafminer, *Leucoptera coffeella.* Gallardo and Wolcott have reported 15 parasitoids from Puerto Rico. Gallardo has proposed a biocontrol program utilising the augmentation method using the parasitoid, Mirax insularis. This idea has extraordinary merit and should be pursued vigorously. There are many other natural enemies reported that could be considered when their efficacy and mass rearing techniques are determined. Foreign exploration is another potential alternative which has not been attempted for this pest, which supposedly originated in Reunion.

Diamondback Moth

The diamondback moth (DBM) *Plutella xylostella* is a lepidopteran pest of importance to the Caribbean with a significant number of natural enemies. It has outstanding potential for a successful biocontrol program. In Puerto Rico several parasitoids are particularly effective. *Diadegma insularis,* may have rates of parasitism up to 90% when protected from pesticides. One limitation is the very low economic injury level of the pest. However, the selection of planting dates, use of selective pesticides and the introduction of the most effective parasitoids will undoubtedly improve the level of biocontrol.

Alam reported complete biocontrol with *Trichogramma* spp., and *Cotesia* sp. Bennett & Yasseen reported effectiveness of *C. plutellae* against DBM in Barbados, Montserrat, St. Vincent and Trinidad.

Corn Earworm

The corn earworm (CEW) *Heliothis sea* is the limiting factor for sweet corn and other vegetable crops. Chemical control is very difficult, even with and effective insecticide. However, many natural enemies have been observed attacking the eggs and larval stages of this insect. The wasp *Trichogramma* sp. and the predator *Orius pumilio* are considered by Figueroa to be the most important natural enemies in Puerto Rico. In the US, CEW infestations are considered more severe in dry than in wet summers because moist conditions favor the development of several fungal and bacterial diseases. Also several species of predators and parasitic wasps attack it. Populations of these natural enemies can significantly reduce or almost eliminate CEW infestations early in the season. Therefore, early season sprays should be delayed in order to allow populations of natural enemies to develop. Frequent sampling and control measures are necessary in vegetables because even minimal damage by this and other pests can render produce unmarketable.

Melonworm

Usually the melonworm *Diaphania hyalinata* (L.) is the most important insect pest of cucurbits in the Caribbean. Medina et al. observed several parasitoids and predators in Puerto Rico. No natural enemies have been studied in detail. In Barbados, Alam reported melonworm egg parasitised by *Trichogramma pretiosum*, the larvae by *Eiphosoma dentator* and by *Cotesia* sp., and the pupae by *Brachymeria* sp. There is a great potential for the biocontrol of this pest, particularly with microbial insecticides, which are already being used in some countries like Guatemala.

Scale Insects, Whiteflies, Aphids and Thrips

The potential for biocontrol of small insects is well known. Since the introduction of *Rodolia cardinalis* in California in 1892 for control of the cottony cushion scale or the biocontrol of the citrus blackfly in the Caribbean, many outstanding examples have occurred. In Florida, a biocontrol project is seeking natural enemies for whiteflies. As a result several natural enemies have been recorded from Florida, California, the Caribbean, and Central and South America.

REFERENCES

Bryan, M. D., R. J. Dysart and T. L. Burger, "Releases of Introduced Parasites of the Alfalfa weevil in the United States, 1957-88", *USDA APHIS Misc. Publ. No. 1504,* 1993.

Caltigirone, L. E., "Landmark examples in classical biological control", *Ann. Rev. Entomol,* 1981.

Day, W. H., "Biological control of the alfalfa weevil in the Northeastern United States", *Chap 25 In. Papavizas, G. C. ed. Biological Control in Crop Protection,* Allanheld, Osmun & Co., Montclair, NJ, 1981.

Debach, P., and D. Rosen, *Biological Control by Natural Enemies,* Cambridge Univ. Press, Cambridge, UK, 1991.

Doutt, R. J., and J. Nakata, "The Rubus leafhopper and its egg parasitoid: an endemic biotic system useful in grape pest management", Environmental Entomology, 1973.

11

Botanical Insecticides

Plants have evolved for over 400 million years and to defend themselves from insect attack they have developed protection mechanisms such as repellents and even insecticidal effects. The oldest known pest control method is human sacrifice, but since it is not effective, or maybe because of the lack of volunteers, dusts began being used and also plant extracts which are mentioned even in the Bible. Massive use of these insecticides has had a long and difficult road because the earliest data gathering done by researchers among farmers and natives revealed a lot of practices based on superstition, which, when tested by scientific methods. were not shown to be effective.

After the Second World War the few plant and plant extracts that had shown promising effects, and were of widespread use, were replaced by synthetic insecticides. When synthetic insecticides appeared in the 1940's some people thought that botanical insecticides would disappear forever but problems like environmental contamination, residues in food and feed and pest resistance brought them back to the fore. There is no doubt botanical insecticides are an interesting alternative to insect pest control, and on the other hand only a few of the more than 250,000 plant species on our planet have been properly evaluated for this purpose. This means the potential for the future may be huge. In fact, plants like neem (*Azadirachta indica* J.,

Meliaceae), have shown excellent results and there already are commercial products in the market made from it. But one should not think success is at hand and botanical insecticides will replace all synthetic products. They are only alternatives that may be used in Integrated Pest Management programs and they should be used together with other available control measures.

Use of plant extracts and powdered plant parts as insecticides goes back at least as far as the Roman Empire. For instance, there are reports that in 400 B.C. during Persian king Xerxes' reign, the delousing procedure for children was with a powder obtained from the dry flowers of a plant known as pyrethrum (*Tanacetum cinerariaefolium*, Compositae). The first botanical insecticide, used as such, dates back to the XVII Century when it was shown that nicotine, obtained from tobacco leaves, would kill plum beetles. Around 1850 a new plant insecticide known as rotenone was introduced. It is obtained from the roots of plants called timbó. Up to that time this plant was used for fishing purposes only as natives had known for a long time that throwing root pieces to the water caused fish to start floating a few minutes later, making them very easy to catch. Later on plants with irritating properties like incense and sabadilla were used: extracts from the latter plant were used as decongestants. These plants did not kill insects directly but it was said that they "scared them off." More recently other plants used are quasia (*Quaisa amara*, Simaroubaceae), neem or Margosa (*A. indica*) mentioned above, which besides giving excellent results for insect control are also a source of compounds used against cancer.

In México and several Central American countries even today it is common practice to treat pests with plants known for their insecticidal properties as far back as the era of the Aztecs and Mayans. A case in point is the use of a mixture of corn and beans with chili peppers (*Capsicum frutescens*; Solanaceae), rue (*Ruta graveolens*; Rutaceae) or garlic (*Allium cepa*; Alliaceae).

At the present time there are a number of botanical insecticides being marketed which are extracted from neem, grapefruit seeds and garlic, among other plants. Besides, there are the synthetic copies of natural active ingredients like neonicotinoids where imidacloprid stands out. Lastly, it is worth mentioning that this is a field where new discoveries are made every day. One example is the development of a new kind of insecticide obtained from the plant of Chilean Ándes origin known as *Calceolaria andina* (Scrophulariaceae).

NATURE OF THE COMPOUNDS

Plants are like natural laboratories where a great number of chemicals are biosynthesized and in fact they may be considered the most important source of chemical compounds there are. Primary plant metabolism synthesizes essential compounds which are present in all plant species. On the other hand, the end products of secondary metabolism are neither essential nor universally present in all plants. Common among these metabolites are componds with protective action against insects, such as alkaloids, non-proteic amino acids, steroids, phenols, flavonoids, glycosids, glucosinolates, quinones, tanins and terpenoids. Some people support the idea that these compounds do not have a clearly known role, and have even classified them as "metabolic garbage." However, other researchers point out that those products are important chemical signals in the ecosystem. There is a great deal of variation as far as the concentration of secondary compounds that individuals of a population can express. Besides, there is neither a pattern of maximum yield nor particular storage organs for secondary metabolites, even though it is common that the highest concentrations of these kinds of compounds are found in flowers and seeds.

By definition, an insecticide is a substance that performs a biocidal action on insects due to the nature of its chemical structure. For example, if we kill an insect for our insect museum using a cyanide killing jar we can say that cyanide has an insecticidal effect. However, we cannot say the same about water if raindrops kill aphids, since that mortality cannot be attributed to water's chemical structure or properties.

Most plant species used for plant protection exhibit an insect deterrent rather than insecticidal effect. This means that in some way those compounds inhibit normal development in insects. This can be done in different ways that will be briefly described below.

Insect Growth Regulators

The growth regulator effect may be seen in several ways. First are those molecules inhibiting metamorphosis, in other words, compounds preventing metamorphosis from taking place at the right time. Other compounds force the insect to go through an early metamorphosis, so that development takes place at a time not favorable for the insect. Lastly, it has been observed that certain molecules may alter hormones related to this function so that insects suffer malformations, are sterile, or die. In fact, one of the most common entomological anecdotes tells that an experiment was being carried out simultaneously in the United States and Hungary. In the latter country the insects went through an additional immature stage before molting to adults. All methods were reviewed but no difference was found until the paper towels used to provide water for the insects were analysed and then it was found that in the two countries the paper towels were made from different trees. In Europe they used *Abies balsamea*, a coniferous tree, very common in Hungary, which contains a plant hormone which induced the additional molt. A more practical example is the case of basil plant (*Ocimum basilicum*) from which the compound juvocineme II is extracted and which later on served as the model for synthesis of piriproxifen and fenoxicarb.

Feeding deterrents

Feeding deterrence is perhaps the most studied mode of action for plant derivatives used for insect pest management. Strictly speaking, a feeding deterrent is a compound that once probed by the insect, causes it to stop feeding and starve to death. Many compounds showing this activity are terpenes and most have been isolated from medicinal plants native to Africa and India.

Repellents

The use of plants as repellents is very old but has not received the necessary attention for proper development. Use has been achieved with compounds having bad odor or irritant effects as is the case for garlic and hot peppers. A clear example can be seen in uses by Guatemalan and Costa Rican natives who used to "paint" containers or sprinkle them with garlic powder before using them to store corn or beans to prevent the grain from being damaged by weevils (weevil punctures) and also to shoo away rodents. Lastly, it is not uncommon to hear of home recipes mentioning the use of fennel (*Foniculum vulgare*), rue (*Ruta graveolens*) and eucalyptus (*Eucaliptus globolus*) among other aromatic plants to repel clothes moths.

Confusants

Chemical compounds in a given plant are unequivocal signs for insects to find their food source. In fact, there are cases like the monarch butterfly, which feeds on plants that are highly poisonous to other organisms, but that are found by the butterfly because of the poisonous substance. One way to use this characteristic in IPM has been setting up traps and either spraying them with infusions of certain plants that are more attractive to the insect or of the same plant but from other areas so that the insect will have so many sources of stimuli that they will not be able to find the plant we wish to protect. Another option is to set up traps containing watery extracts of the plant so that insects will "land" in the traps and not on the crop.

So, thinking about the above mentioned information we should consider all those compounds which we know have insect deterrents and have preventive rather than curative effects.

Advantages and Disadvantages of Botanical Insecticides

Advantages

1. Plants producing the above mentioned compounds are known by the farmer because most of the time they grow in the same general area.
2. Often these plants also have other uses like household insect repellents or are plants with medicinal applications.
3. The rapid degradation of the active product may be convenient as it reduces the risk of residues on food.
4. Some of these products may be used shortly before harvesting.
5. Many of these products act very quickly inhibiting insect feeding even though long term they do (will?) not cause insect death.
6. Since most of these products have a stomach action and are rapidly decomposed they may be more selective to insect pests and less aggressive with natural enemies.
7. Most of these compounds are not phytotoxic.
8. Resistance to these compounds is not developed as quickly as with synthetic insecticides.

Disadvantages

1. Most of these products are not truly insecticides since many are merely insect deterrents and their effect is slow.
2. They are rapidly degraded by UV light so that their residual action is short.
3. Not all plant insecticides are less toxic to other animals than the synthetic ones.
4. They are not necessarily available season long.
5. Most of them have no established residue tolerances.
6. There are no legal registrations establishing their use.
7. Not all recommendations followed by growers have been scientifically verified.

INSECTICIDAL PLANTS

There are a several plants with insecticidal properties. For example, in 1950, Heal *et al.* reported approximately 2,500 plants in 247 families with some sort of toxic property against insects. But to use them it is not enough that the plant be considered as promising or even with proven insecticidal properties. It is also necessary to conduct an analysis of the risks to the environment and to human health. Another example is that it may not be appropriate to recommend the use of plants in danger of extinction, that are difficult to find, or if their use results in important alterations to their population density in natural conditions. For example, if tomorrow it is discovered that baoba wood has a compound which may be an insecticide, that does not mean we are going to cut the baoba trees. For these reasons and with the purpose of obtaining maximum benefit from a plant with insecticidal properties, but without degrading the ecosystem, the properties that an ideal insecticidal plant should have are listed below:

1. It should be a perennial.
2. It should have a wide distribution and be present in large numbers in nature. Otherwise it should be possible to be grown by agricultural procedures.
3. The plant parts to be used should be removable: leaves, flowers or fruit.
4. Harvesting should not mean destruction of the plant (avoid the use of roots or bark).
5. The plants should require small space, reduced management and little water and fertilisation.
6. The plant should have additional uses (for example medicinal use).
7. The plant should not otherwise have a high economic value.
8. The active ingredient should be effective at low rates.

It would be a big mistake to consider products of plant origin, and this includes botanical insecticides, harmless merely because they are natural. There are large numbers of plant products which are highly toxic; remember that ancient history mentions that Socrates was sentenced to death by drinking a diluted extract of hemlock (*Cicuta* spp.). Schmutz and Breazeale, in their book "Plants that Poison" point out around 120 species of plants containing substances that are toxic to humans, including plants as common as almond trees, some bean species, garlic, strawberries and apples, among others. So we must not forget that the toxic potential of a molecule is the nature of its chemical structure and not its origin. As Paracelsus wrote in 1564 "the difference between something that kills and something that cures is the rate."

Plants have been used not only directly as insecticides but also as a source for a series of synthetic insecticides developed in the laboratory. From the pest control point of view. one problem is that botanical insecticides disappear soon after application because they are decomposed by the action of light and temperature so that they remain on the plant only briefly (typically no more than 24 hours). This has prompted industry to modify molecules found in plants so that they are more toxic and more persistent.. These improvements offer the advantage that the crop does not have to be sprayed virtually every day and that there will be less pressure to harvest the insecticidal plants into extinction. A clear example of this may be found in two insecticide families of widespread use in agriculture, medicine and urban environments: pyrethroids and carbamates which are synthetic derivatives of molecules isolated from plants like pyrethrum (*T. cinerariaefolium*) and Calabar beans (*Physostigma venenosum*), respectively.

DESCRIPTION OF SOME COMPOUNDS

Rotenone

Rotenone is a flavonoid extracted from the roots of two plants: *Derris* spp. (Fabaceae) and *Lonchocarpus* spp. (Fabaceae). The first one gives up to 13% of rotenone while the second only about 5%. *Derris* spp. is a native to Eastern tropics, while *Lonchocarpus* spp. is native to western hemisphere. Rotenone is a contact and ingestion compound, which acts as a repellent too. Its mode of action involves the inhibition of the electron transport at the mitochondrial level, thus blocking phosphorylation of ADP to ATPthereby inhibiting insect metabolism. Insects poisoned with Rotenone show the following symptoms of intoxication: a drop in oxygen consumption, respiratory depression and ataxia leading to convulsions and finally to paralysis and death by respiratory arrest.

Sabadilla

This is a compound that can be obtained from seeds of a plant of South American origin known as *Schoenocaulon officinale* (Liliaceae). Seeds of this plant have been shown to have high concentrations of alkaloids which impart its toxic properties. The mode of action is disruption of neuron cell membranes causing reduction of nerve activity, paralysis and death. The ground seeds are one of the plant insecticides with the lowest mammal toxicity but that is not the case with their isolated alkaloids which are both highly toxic and skin irritants.

Nicotine

Nicotine is an alkaloid obtained from some plants in the Solanaceae family, particularly tobacco (*Nicotiana tabacum*). The insecticidal properties of nicotine were recognised in the first half of the XVI century. This compound is not found free in the plant but in the form of maleates and citrates. Nicotine is essentially a non-persistent contact insecticide. Its mode of action consists in mimicking acetylcholine when it binds with its receptor in the post-synaptic membrane of the muscular union. The acetylcholinic receptor, is a site of action of the postsynaptic membrane which reacts with acetylcholine and alters the membrane permeability. Nicotine activity causes the production of new nerve impulses which cause convulsions, and death. An important new class of insecticides, commonly known as neo-nicotinoids, are synthetic copies or derivatives of the nicotine structure. These include imidacloprid, thiacloprid, nitempiram, acetamiprid and thiamethoxam, among others.

Ryania

This compound is obtained from the roots and stems of a plant native to South America known as *Ryania speciosa* (Flacourtiaceae). From this plant may be obtained a series of alkaloids, of which the most important is ryanodina. This alkaloid is effective as a contact or stomach poison and directly prevents muscles from contraction,causing paralysis.

Azadiractin

This compound is a tetraterpenoid characteristic of the Meliaceae family but particularly from the Neem tree (*A. indica*), indigenous to India. The compound is found in bark, leaves and fruits of the tree but seeds have the highest concentration. This compound has not yet been synthesized in the laboratory, but when isolated and tested pure the results have been less than when extracts are used. In the extract 18 compounds have been identified among which salanine, meliantrol and azadiractin are most prominent, the latter being in the highest concentration. Azadirachtin shows antifeedant activity, is a growth regulator, inhibits oviposition and is also a sterilizing compound. Today, commercial formulations of neem may be found with names like Neem Gold, Neemazal, Econeem, Neemark, Neemcure and Azatin among others, in many countries including the United States, India, Germany and several Latin American countries.

Pyrethrins

Pyrethrins, which are esters with insecticidal properties, are obtained from pyrethrum (*T. cinerariaefolium*) flowers. The compounds obtained from this plant which have known insecticidal activity are six esters formed by the combination of the acids chrysanthemic and pyrethric and the alcohols piretrolone, cinerolone and jasmolone. These compounds act both on the central nervous system and in the peripheral nervous system causing repetitive discharges, followed by convulsions. Research has shown that these compounds block sodium ion influx resulting in the channels being affected by intermolecular forces which causes alterations in moving ion conductivity as a result of tchanges in the channels. There is no doubt that the most important characteristic of these compounds is their irritating effect or "knock down" which causes the insect to stop feeding as soon as it encounters a treated surface.

Pyrethrins are the best example of products copied in the laboratory since their modifications gave rise to the pyrethroid family of insecticides.

Resistance

Most insecticides of botanical origin are extracts containing a group of active ingredients of diverse chemical nature. From a resistance point of view the short residual life of plant insecticides may be considered as a positive since there will be a very low probability that two extracts would always be identical so that selective pressure on the pest species will not always be the same. Even if all the same compounds are found in the extract, concentrations almost always will be different. Generally speaking, insect resistance takes longer to develop to a mixture of natural active compounds than to any one individual component. This may be because it is more difficult to detoxify a compound complex than a single molecule. In a lab evaluation with green peach aphid, *Myzus persicae,* when azadirachtin was applied alone, the insect was able to develop a level of resistance nine times higher than in the initial population within 35 generations. Using the whole neem extract (containing the same azadirachtin concentration) no indication of resistance was seen in the same time period.

FUTURE PERSPECTIVE

The main market for plant origin insecticides is in parks and gardens. This is due to their low environmental persistence so that people will be less exposed to the toxic compounds. It is expected that within 10-15 years, these compounds will increase 25% in insecticide market share and they will not be limited to garden areas but there may be a massive growth towards urban and agricultural uses. However, even though plant origin insecticides are advantageous options from an ecological point of view, it would be unreasonable to think they will completely replace synthetic insecticides.

On the contrary, it will be quite logical to expect a complementary use and the coexistence of the two kinds of compounds as it happens today with pyrethrum and synthetic pyrethroids in some IPM programs. On the other hand, organic agriculture is a market with high demand for plant origin insecticides as organic growers cannot use conventional agrichemicals. At this time this market is expanding and as a rule rates of return are high, so that this is an important "niche" to take care of. Unfortunately it is commonly thought that organic agriculture is the same as conventional agriculture except that no synthetic agrichemicals are used. This mistake may lead some growers to lose confidence in their objectives if they do not get the expected results and they may hold on to their old preference for synthetic insecticides.

Plant origin insecticides also have a great advantage by being compatible with other low risk options which are acceptable for insect management, such as pheromones, oils, detergents, entomopathogenic fungi, predators and parasitoids, among others, which greatly increase probabilities of being integrated in IPM programs. Lastly, it may be pointed out that, short term, there is no doubt that new plants will be studied, as is the case of *Calceolaria andina*. New technologies will be developed and perfected to isolate and identify molecules. Extension and education programs on plant origin insecticides will increase and improve and new sources of raw materials will be found for their manufacture, as it is happening today with discarded citrus peels in the United States.

REFERENCES

Addor, R. W., "Insecticides", In: Godfrey C.R.A. (Ed) *Agrochemicals from natural products,* Marcel Dekker, Inc., 1995.

Arnason, J. T., B. J. Philogene and P. Morand (EDS), *Insecticides of plant origin,* American Chemical Society, Washington, DC. USA, 1989.

Coats, J. R., "Risks from natural versus synthetic insecticides", *Annu. Rev. Entomol,* 1994.

12

Cultural Control of Insect Pests

"Cultural control" is the deliberate alteration of the production system, either the cropping system itself or specific crop production practices, to reduce pest populations or avoid pest injury to crops.

FUNCTIONAL MECHANISMS

1) impediments to pest colonisation of the crop
2) the creation of adverse biotic conditions that reduce survival of individuals or populations of the pest
3) modifications of the crop in such a way that pest infestation results in reduced injury to the crop
4) enhancement of natural enemies by manipulating the environment

Destruction or Provision

Breeding or Overwintering Refugia

Many natural enemy species require food sources in the form of pollen, nectar, or innocuous arthropods that are not present in particular crop habitats. These food requirements may be provided to support natural enemy populations by encouraging or deliberate development of certain wild vegetation habitats near plantings of the crop.

Sometimes a second species of prey may be essential for a natural enemy to survive in a region from year to year. One of the better documented examples is based on research by Doutt and Nakata with Anagrus epos, an egg parasite of the grape leafhopper, Erythroneura elegantula. The grape leafhopper is the major pest of grapes in the San Joaquin Valley of California. This wasp overwinters in the egg stage and the grape leafhopper overwinters as adults.

The wasp overwinters in the eggs of the blackberry leafhopper, Dikrella californica, a noneconomic species whose eggs are present throughout the year on wild blackberries. Overwintering wasp populations tend to be largest in blackberries along rivers that have an overstory of sheltering trees. When the blackberries produce new foliage in February, heavy oviposition by Dikrella leafhoppers is stimulated. The parasite population increases enormously on the eggs so that by late March and early April there is widespread dispersal of newly-produced Anagrus wasps. Doutt et al. reported that vineyards located within 5.6 km of an established blackberry refuge will benefit immediately from the immigrant parasites. Vineyards beyond this distance rarely were colonised by wasps until mid-season.

It was presumed that all one needed to do was to establish a blackberry planting adjacent to the vineyard to support non-pest populations of leafhoppers and wasps. However, because of the lack of sheltering trees, as occurs in natural blackberry systems along rivers and streams, populations of Anagrus never became permanently established. These blackberries were being irrigated on the same regime as the vineyard and blackberry plants require more moisture than the grapes. This resulted in the *Dikrella* populations dying-out. *R.* ursinus, California blackberry, is a native species which grows best in shady, moist conditions, while *R. procerus*, Himalaya blackberry, is endemic to Europe and grows well in both the shade and sun. When these plants were grown under shade with high moisture they produced a sufficient population of *Dikrella* for maintaining *Anagrus* populations.

Kido et al. presented data showing a correlation between adult *Anagrus* dispersion in early spring from French prune orchards and parasitism of grape leafhopper in nearby vineyards. *Anagrus epos* was parasitising another leafhopper species, *Edwardsiana prunicola* which feeds on prune. Both blackberry and French prune provide alternate hosts for the wasp, and effectively are nurseries for producing *Anagrus* wasps.

Most aphid species in the temperate areas of North America overwinter as eggs on their primary woody host. The green peach aphid (GPA), *Myzus persicae*, overwinters on peach and wild cherry and a number of other *Prunus* spp. Green peach aphid vectors maize dwarf mosaic virus of sweet corn. Because GPA does not migrate into horticultural crops until early July, it is primarily the later plantings of sweet corn that show symptoms and yield reductions. We recommend to our growers to plant the early plantings of sweet corn near to peach trees, so that the later plantings can escape heavy probing by these vectors as they move from their primary host plants. A similar situation exists with GPA as a vector of cucumber mosaic virus of green pepper. Again, either the destruction of Prunus spp. or planting away from these primary hosts can reduce losses to these viruses.

Alternate Hosts or Volunteer Plants

The black cutworm, *Agrotis ipsilon*, is a major pest of corn seedlings in the corn belt states, especially in a no-till system. Young larvae feed on seedling corn leaves until fourth instar when they cause serious damage by cutting or drilling the plants. However, if the seedlings can reach the four leaf stage before being infested, no significant yield reductions occur, creating a "the two-edged sword."

However, if the grower waits until the corn reaches the four leaf stage before cultivating or using herbicides to control the weeds, yield reductions occur due to weed competition. So, the best management tactic is to time preplant herbicides at least 14 days before planting to minimise ovipositional sites and early instar food sources.

Colorado potato beetle, *Leptinotarsa decemlineata*, overwinters as adults in fields of origin or in the undisturbed habitats adjacent to fields. The beetle emerges in the spring and it must regenerate its flight muscles before it migrates. If a host plant is not within walking distance of the beetle, hence no food source, it can fly several kilometers in search of a host. However, if it emerges into a host habitat, such as a potato field, it will feed and oviposit, and after about a week the beetle will fly from the field to colonise another host habitat. If volunteer plants are allowed to grow in the previous years field, the emerging beetles will colonise these plants, feed and oviposit before migrating, hence, delaying the colonisation process of adjacent fields.

Crop Rotation or Maintenance of a Host-free Season

Crop rotation interrupts normal life cycle of insect pests by placing the insects in a non-host habitat. Rotation is generally most successful against arthropod pest species with long generation cycles and with limited dispersal capabilities.

White fringed weevil complex, Graphognathus leucoloma (Boheman) and *G. peregrinus* (Buchanan), adults lay many eggs when fed on soybean and cause heavy damage to this crop. However, the grass crops, including corn, are in some way nutritionally deficient to support feeding, and do not suffer damage from this pest. So, a soybean/corn rotation is effective and economical.

Crop rotation interrupts the life cycle of the northern (*Diabrotica longicornis*) and western (*D. virgifera*) corn rootworms, and both species can be effectively controlled by rotation throughout the corn belt of the U.S. This tactic has been compromised in some areas where the rootworms diapause for more than one year.

Overwintered Colorado potato beetle (CPB) can disperse to colonise new fields; however, after they emerge from the soil they need to regenerate their flight muscles and they will not oviposit until they have fed on a nutritionally acceptable host plant. So, between the colonisation process and physiological development, a rotated field that was rotated as little as 200m from last years field can be colonised 1-2 weeks later and generally at lower population densities than a non-rotated field. This saves 1-2 sprays for first generation larvae. And, often pushes most emergence of summer adults after August 1 (diapause switch). Those beetles that emerge after diapause induction do not produce flight muscles or a reproductive system. When rotated with corn (sweet or field) where large populations of *Coleomegilla maculata* (CMAC) buildup feeding on pollen and aphids, CMAC move into the potato fields in late spring or early summer to feed on CPB eggs.

Tillage

Tillage operations used to produce a crop include soil turning and residue-burying practices, seedbed preparation, and cultivation. Some forms of tillage can reduce pest populations indirectly by destroying wild vegetation (weeds) and volunteer crop plants in and around crop-production habitats.

Overwintering populations of *Heliothis zea* may be greatly reduced by either fall or spring plowing operations.

Overwintering survival of the soybean stem borer *Dectes texanus* is inversely related to depth of burial of soybean crop residue following harvest.

The stalk borer, *Papaipema nebris,* has become a major pest of corn in the midwest where farmers use conservation tillage. That is, serious infestations have occurred throughout entire fields where no-till is practiced. Using a fall moldboard plow and a spring disk reduced borer damage when compared to no till. However, other studies suggest that stalk borers would be better managed by controlling grassy weeds within fields in the late summer and early fall to prevent oviposition rather than relying on tillage or weed control practices to reduce populations of eggs and larvae after oviposition has already taken place.

Timing of planting or harvest

Alterations in planting date and harvest date can frequently result in plants escaping from damaging pest infestations.

Delayed planting

Late planted potatoes (>June 15 emergence) in Massachusetts in non-rotated fields suffer less damage than conventional planting dates (plant emergence prior to June 15). Overwintered beetles remain in the field for about 5-7 days after emerging from the soil, and if no host plants are present, they leave the field by flight. 90% of the overwintered population emerges by early June, and a late planted field is only colonised by a small proportion of this population. Because oviposition occurs late, first generation adults do not emerge until after the induction of diapause, resulting in only one generation of larvae to control.

Buntin et al. showed that by delaying the planting date of wheat in the fall wheat seedlings were not present for oviposition by the Hessian fly, *Mayetiola* destructor. Infestations of fall and winter wheat by this pest, declined without enhancing spring infestations or reducing wheat yields.

Early harvest date

Pink bollworm, *Pectinophora gossypiela*, overwinters as last instar larvae and diapause is controlled by short days (<13 hours of light). By harvesting early (before diapause induction) the number of overwintering larvae is reduced to low levels. Practices: 1) defoliate or desiccate the mature crop to cause all bolls to open at nearly the same time, expediting machine harvesting; 2) harvest the crop early, shred stalks, and plow under crop remnants immediately; 3) irrigate prior to planting in desert areas if water is available to encourage plant growth; 4) delayed planting - plant new crops during a designated planting period, which allows for maximum suicidal emergence of moths that emerge during the spring time, that is, moths emerge and die before cotton fruit is available for oviposition. Short season cultivars are key to this program.

Trap crops

Crop monocultures are often damaged more severely by pests than is the same crop located in an area with crop diversity. However, there are cases where such diversity can aggravate pest problems. It is in these situations where trap crops can be important.

Lygus bugs are a key pest of cotton in the San Joaquin Valley of California. A major habitat of the bugs is alfalfa fields, which are commonly interspersed with cotton fields. When the alfalfa is harvested the bugs leave the field in large numbers to infest cotton fields. By strip cutting the alfalfa field, the number of bugs migrating to cotton can be minimised. A twist on this approach is to plant strips of alfalfa every 400' between cotton fields and then only harvest 1/2 of the alfalfa strip at any one time. Stern demonstrated that *Lygus* bugs concentrate in these strips, and the need for insecticides was virtually negated. Furthermore, beneficial parasitoids and predators were extremely abundant in these strips and, as they moved to and from the adjacent cotton, an added benefit resulted.

A large proportion (>70%) of Colorado potato beetles overwinter in uncultivated habitats adjacent to potato fields. Movement from these overwintering sites by beetles to colonise potato fields can be inhibited by using plastic-lined trenches around the perimeter of the field, or by treating the outer 6 meters of the field with an insecticide that kills the colonising beetles. The perimeter plants effectively act as a trap crop, or barrier. The plastic-lined trench has been shown to be 84% effective at controlling these beetles, while fields treated with the newly registered insecticide, imidacloprid, controlled 100% of the beetles. Because the colonising population has been reduced to low densities, it is then possible to use the *Bacillus thuringiensis tenebrionis*-based insecticides to manage the larvae feeding within the interior of the field.

Sanitation

Destroying culled potatoes by turning them under, so that they decompose is important in reducing inoculum sources of the late blight fungus, and as an early season host for aphids and other pests.

Destroying slash in logging operations can reduce the buildup of bark beetles.

Water or nutrient management

Water can be used directly to suffocate insects or indirectly by changing the overall health of the plant, while fertiliser can influence the injury to a crop primarily through alterations in crop growth or nutritional value to the pest. Some pest populations are enhanced by poor crop growth, while others are enhanced by succulent crop growth.

Water

Flooding of cranberry bogs is a valuable tool for controlling several insect pests (mites and fireworm) and plant pathogens of cranberry. Flood irrigation is frequently used to reduce populations of wireworms in vegetable and sugarcane crops. Likewise, flooding can be used to control white grubs in sugarcane, especially under conditions of high temperature.

Furrow irrigated potato fields in many parts of California tend to crack upon drying, exposing potato tubers to ovipositing potato tuberworms. In areas where this is a problem, overhead sprinkler irrigation is recommended to prevent the cracking of soil.

Overhead sprinkler irrigation can enhance dissemination and infectivity of some entomopathogenic organisms, especially fungal pathogens. Such practices help to promote epizootics of *Nomuraea rileyi* in populations of the velvetbean caterpillar in soybean.

Fertilisation

Enhancement of succulent cotton growth through fertilisation renders the crop more attractive to populations of the cotton aphid, *Aphis gossypii*, cotton fleahopper, Pseudatamoscelis seriatus, and the cotton bollworm. *Heliothis zea.*

Undernourished plants are often more attractive to colonising aphids because they are more yellow and reflect more light in 540nm range.

Physical barriers

Floating row covers were used to protect broccoli plants from the imported cabbageworm, *Artogeia rapae* (L.), flea beetle complex, striped cucumber beetle, *Acalymma vittata* (F.) and cabbage maggot, *Delia radicum* (L.). The row covers were very effective at protecting the plants from all pests except the cabbage maggot. The cabbage maggot overwinters as pupae in the soil, so when the row covers were placed over the soil the flies emerged and became trapped underneath the cover.

Floating row covers were placed over potato plants being grown for seed to prevent aphids from vectoring PLRV (persistent virus) and PVY (non-persistent virus). The covers were more effective at preventing transmission of PLRV than PVY, mainly because it takes prolonged feeding by the aphid to vector the persistent virus PLRV, while PVY can be vectored within 30 seconds.

Border sprays

This really isn't a cultural control strategy *per se*. However, the success of using border sprays is based on knowing how an insect colonises and moves within a cropping system.

Chouinard et al. targeted sprays to the outer row of apple trees to protect the fruit from the plum curculio, *Conotrachelus nenuphar* (Herbst). This pest overwinters outside of the orchard and migrates into the orchard at petal fall. So, an application at petal fall and another at pink stage reduced the damage at harvest from this pest from 59% without any sprays (targeted against this pest) to 2.4%.

Cold/Heat Treatments

Grain chilling is a non-traditional, non-chemical preservation technology for the storage of cereal grains. A grain chiller utilises a refrigeration system to control the temperature and moisture content in stored grain independent of ambient conditions. Over 1 billion bushels of grain are chilled annually in countries like Argentina, Australia, Germany, Great Britain, France, Indonesia, Israel and Mexico, but very little within the U.S. grain industry. It costs about 0.5 to 1 cent per bushel (insecticide costs 0.33 to 1 cent). At temperatures below 65°F, insects and molds cannot survive, or at least their rate of development is reduced. Cool air from outside is passed over a bank of refrigeration coils in order to decrease the air temperature, and the air is then reheated a few degrees to reduce the humidity before it is circulated through the grain.

REFERENCES

Armbrust, E. J. and G. G. Gyrisco, "Forage crops insect pest management", In R. L. Metcalf and W. H. Luckmann, *Introduction to Insect Pest Management*, John Wiley and Sons, New York, 1975.

Buntin, G. D. , P. L. Bruckner and J. W. Johnson, "Management of Hessian fly (Diptera: Cecidomyiidae) in Georgia by delayed planting of winter wheat", *Journal of Economic Entomology*, 1990.

Engelken, L. K., W. B. Showers and S. E. Taylor, "Weed management to minimize black cutworm (Lepidoptera: Noctuidae) damage in no-till corn", *Journal of Economic Entomology*, 1990.

13

Sterile Insect Release Method

INDUCED STERILITY IN INSECTS

Discoveries in such diverse disciplines of science as radiation biology, chemistry, animal behavior, molecular biology, genetics and entomology have contributed to our current ideas concerning the control of insect pests. In 1895 W. K. Röntgen observed and reported his discovery of X-rays. By 1903 the effects of such rays from radium upon the reproductive systems and development of insects had been reported by C. Bohn. G. A. Runner, in 1916, found that low doses of X-rays decreased the reproductive capacity of the cigarette beetle, while higher doses killed them.

By 1926 H.J. Muller had demonstrated that X-ray radiation will cause heritable changes in fruit flies (*Drosophila melanogaster* Meigen). Dominant lethal mutations were found to be one of the most frequent types of induced mutation. The "partial sterility" of X-ray treated males was of concern to Muller because this effect reduced the numbers of visible mutations which could be recovered for a given dose of radiation.

As early as 1937, E. F. Knipling had conceived an approach to insect control in which the natural reproductive processes of the screwworm fly (*Cochliomyia hominivorax* Coquerel) are disrupted by chemical or physical mechanisms thus rendering the insects sterile.

These sterile insects are then released into the environment in very large numbers (10 to 100 times the number of native insects) in order to mate with the native insects that are present in the environment. A native female that mates with a sterile male will produce eggs, but the eggs will not hatch (the same effect will occur for the reciprocal cross). Since there are 10 to 100 more sterile insects in the population than native insects, most of the crosses are sterile. As time goes on, the number of native insects decreases and the ratio of sterile to normal insects increases, thus driving the native population to extinction.

This unique insect control method is now known as the sterile insect technique (SIT), or the sterile insect release method (SIRM). SIT proved to be an extremely successful method for control of the screwworm fly when ionising radiation was used as the sterilising source. The success of the screwworm SIT program led to investigations of the effects of radiation on the reproductive performance of many other economically important insects.

The use of mutagenic chemicals to sterilise insects for sterile insect release programs was also considered because this seemed to be an efficient method of causing sterility. For example, chemical sterilants could be added to the rearing diet, or the chemical could be applied to eggs or pupae at times when they are being handled in the normal rearing process. However, the use of chemical sterilants is presently very limited because of fears of environmental contamination by insects carrying carcinogenic materials. There are also problems with disposing of the diet or other carriers used to administer the chemosterilant to the insects. Certain types of photo-chemically reactive dyes have been demonstrated to have insecticidal properties but are environmentally safe. There is a challenge for investigators to discover similar chemicals that could be used as sterilants.

The methods that are presently employed for the successful use of the sterility principle have not changed significantly since E. F. Knipling's original formulation:

1. Techniques that make it possible to produce large numbers of the target insect.
2. Techniques that make it possible to sterilise large numbers of the target insect.
3. Reasonably competitive insects that can be released after sterilisation.
4. Economical systems for releasing and distributing the insects over the treatment area. (Release component)
5. Tools that will assess native populations accurately before and after the release of the treated insects. (Evaluation component)
6. A treatment area large enough (or well-isolated enough) to exclude the possibility of immigration of inseminated females into the release area.

There are difficulties encountered when SIT is applied to certain Lepidoptera, Coleoptera, or to insects (like mosquitoes and houseflies) whose obnoxious behavior precludes the release of large numbers in the vicinity of people or animals. For example, a dose of 5500 rad (55 Gy) is sufficient to guarantee 100% sterility of screwworm flies. However, even after doses exceeding 50,000 rad (500 Gy), some lepidopteran pests will still produce F1 progeny. Investigators have found that these F1 progeny exhibit levels of sterility equal to or higher than those of their treated parents. This inherited sterility has been suggested to be of use in control programs.

The existence of insect species that do not respond to our attempts at sterilisation has led investigators to search for other methods which use genetic principles, but do not necessarily involve radiation-induced sterility. These other "autocidal" control methods are discussed in more detail after we discuss the cases where induced sterility has been used.

Insect Control Programmes

The successful use of sterile insects for the eradication of the screwworm from the United States and Mexico has been detailed in a number of symposia and books. The U. S. D. A., Animal and Plant Health Inspection Service maintains a home page on the screwworm eradication program. The screwworm program has had its share of skeptics and detractors who theorised that the sterile insect release method could never work.

In spite of all the scientific criticism, eradication proceeded quickly across the southern tier of the United States. The US portion of the screwworm eradication program started in Florida in 1957 and by 1966 all self-sustaining screwworm colonies in the US were eliminated. However re-infestations from flies migrating from Mexico compromised the program, so in 1972 a joint United States-Mexico program was initiated. This program enabled Mexico to be officially declared free of screwworms in 1991; Belize and Guatemala in 1994, and El Salvador in 1995. Honduras is considered technically free of screwworms since no flies have been detected since January, 1995. The ultimate goal of a proposed United States-Central America project is to maintain a sterile insect barrier at the Darien Gap in Panama starting in 1997. This program has saved billions of dollars in livestock and wildlife loses since its inception.

Although the screwworm eradication program is the most visible and successful use of SIT, a number of other insect species have been subjected to the release of sterile insects with varying success. This includes the following pests and areas where programs have been carried out: (1) screwworm fly /USA, Mexico, Libya; (2) Mediterranean Fruit Fly (*Ceratitis capitata* Wiedemann)/USA (California),

Mexico; (3) Melon Fly (*Dacus cucurbitae* Coquillett)/Japan, Taiwan; (4) Pink Bollworm (*Pectinophora gossypiella* Saunders)/USA (California); (5) Tsetse Fly (*Glossina* species)/Tanzania, Zimbabwe, Upper Volta; (6) Mosquitoes (various) USA (Florida), East Africa, Venezuela; (7) Boll Weevil (*Anthonomus grandis* Boheman)/ Southeastern USA; (8) Mexican Fruit Fly, (*Anastrepha ludens* Loew)/USA (Texas), Mexico; (9) Gypsy Moth (*Lymantria dispar* Linnaeus)/Northeastern USA, Canada; (10) Stable Fly (*Stomoxys calcitrans* Linnaeus)/USA (St. Croix, Virgin Islands - experimental); (11) Horn Fly (*Haematobia irritans* Linnaeus)/USA (Texas - experimental); (12) Corn Earworm (*Helicoverpa zea* Boddie)/USA (St. Croix, VI); (13) Tobacco Hornworm (*Manduca sexta* Linnaeus)/USA (St. Croix, VI).

One of the more successful and long-term sterile insect release programs involves the pink bollworm in the San Joaquin Valley of California. Over a 24 year period there has been a continuous release of sterile pink bollworm adults during each day of the cotton growing season. This program was initiated in 1968 to protect the *ca* 500,000 ha of cotton in the valley from infestation by the pink bollworm. In other words, this program was conceived as a barrier against the high populations of pink bollworms found in Arizona, Southern California, and Northern Mexico during the decades of 1950 and 1960.

The pink bollworm program has been a cooperative effort of the United States Department of Agriculture, the California Department of Food and Agriculture and the California Cotton Pest Control Board. From 1970 to 1991 the pink bollworm rearing facility in Phoenix produced from 99 million to 826 million dye-marked moths per year. Most of the moths were radiation sterilised and released from airplanes over cotton natives in the San Joaquin; however, large numbers of moths have also been provided for research programs on sterility, insecticide evaluation, and pheromone isolation and identification.

During the course of this release program, the numbers of native moths captured in the San Joaquin Valley has varied considerably from year to year. Because of careful monitoring of the dye-marked sterile and native populations through the use of pheromone baited traps, the program was able to quickly adjust both the patterns of release and numbers of sterile insects released in a given area. Each year releases are made in sections of cotton which had native moth captures the previous year. As native pink bollworms appear in trap captures, the release patterns and rates are adjusted accordingly. In only one year, out of 24 years of the program, have conventional insecticides been used to control pink bollworm infestations and then on a very limited basis. No carry-over of this population from 1990 to 1991 was apparent, and overall numbers of native captures in 1992 were low. To the present time, no other detectable economically damaging populations of pink bollworm have developed in the San Joaquin Valley.

AUTOCIDAL INSECT CONTROL TECHNIQUES

Some species of insects have not proved to be good candidates for the sterile insect technique, so other methods of using the insect to control itself (autocidal control) have been investigated. The following approaches have been investigated.

Inherited Sterility

This effect has also been referred to as: delayed sterility, F_1 sterility, partial sterility, etc. This phenomenon is of use in organisms, such as Lepidoptera and Homoptera, that contain polycentric chromosomes. It involves the transmission of aberrant chromosomes (usually in the form of translocations) from the released population to the native population.

This can be accomplished in two ways: 1) by treating a laboratory-reared population of insects with a dose of radiation which is sufficient to break chromosomes, but low enough to produce little direct sterility, or 2) by culturing stocks of the target species which contains homozygous translocations. When the aberrant chromosomes are passed on to the native populations, the individuals carrying the aberrations in the heterozygous state will show sterility. The amount of sterility will depend on the number and size of the translocations carried by each individual. Some advantages of this method include: 1) enhanced reproductive competitiveness of the partially sterile released individuals compared to the fully sterilised individuals used in SIRM; 2) the F1 individuals are produced in the native population, thus freeing rearing facilities for the production of higher numbers of the primary release organism.

Conditional Lethal Mutations

In this technique, strains of insect are produced (by genetic manipulation) carrying traits that are detrimental to the species in the native environment, but which are not detrimental under laboratory conditions. For example, the inability to diapause would be a conditional lethal in an insect that had to go into diapause to survive a host-free period, but in the laboratory no diapause would be necessary. Some other types of conditional lethal traits which could be used are cold-sensitive lethal mutations, high-temperature-sensitive lethal mutations, inability to fly, lack of sex pheromone production or response, lack of ability to develop on a particular host, change in protective coloration, or any other genetic change which could affect an individual's ability to survive or reproduce in the native. In the past, the introduction of conditional lethals was dependent upon our ability to identify and isolate induced or naturally occurring mutations in the target species.

Investigators are presently investigating the possibility that such mutations can be taken from genetically well known organisms, such as *D. melanogaster,* and introduced into economically important organisms using new techniques developed in molecular biology.

Behavioral Changes

It would be advantageous to cause changes in the specific insect behaviors that make them undesirable or destructive. For example, in many pest species, a change from multivoltine to univoltine development would eliminate the economic destructiveness of the pest. Such a change is often shown to be genetic in character. If we could learn enough about insect physiology to be able to identify the genetic determinants behind such behavior, then we should be able to genetically manipulate the species and release the new strain in such large numbers that the new trait would become predominant in the population. Some other such traits would be alterations of host range, time of development, fecundity, sex-ratio, humidity requirements, pupation sites, etc. We should also consider traits in insects that would make the insect susceptible to cheaper or environmentally benign chemicals.

Hybrid Sterility

The evolutionary background of many insect pests leads us to believe that reproductive barriers may exist between races of the same species in different geographic areas or between closely related species. Sterile hybrids could be raised in the laboratory and released into native populations. This phenomenon has been clearly demonstrated in crosses between *Heliothis virescens* males and *Heliothis subflexa* females and should be more widely investigated.

Simply Inherited Mutations

Genetic changes in a single gene could lead to decreased fitness in the native if the gene could be forced into the native population in large numbers. Some examples of such changes would be recessive lethal mutations, eye color changes (to increase or decrease light sensitivity), pupal or adult body color changes, antennae, leg, or wing deformities.

It is obvious that each of the previously mentioned phenomena requires significant research data on insect genetics, physiology, behavior, and reproduction to bring any of the techniques to fruition.

GENETIC SEXING TECHNIQUES

In many applications of autocidal control, it would be efficient to separate the males and females before release. Possible reasons for such separation are to avoid assortative mating; to avoid any increase in the size of the feral population during a genetic control procedure; to eliminate females which may be disease vectors, extremely obnoxious, or which cause damage to produce or livestock. Another reason for using a genetic sexing procedure is to bring about cost savings in the rearing process if one sex could be eliminated in the egg stage (pre-zygotic sexing). In this case, twice as many insects (of one sex) could be produced with a given expenditure for diet and labor.

Thus, removal of females during the rearing process has been a goal in autocidal control research for many pest insects. The most successful efforts have been with a mosquito, *Anopheles albimanus*, the medfly, *Ceratitis capitata*, and the stable fly, *Stomoxys calcitrans*. Two of these approaches demonstrate what can be done. A combination of mechanical separation of the sexes and genetic manipulation has been employed in the medfly. Wild-type pupal color is brown. Mutants involving either black pupae or white pupae have been linked with the sex chromosomes by means of single or multiple chromosomal translocations. By this means strains are produced in which male pupae are black and female pupae are brown. The pupae are then separated by electronic color recognition circuitry in a mechanical or air-driven sorter. Some disadvantages of this technique are the possible breakdown of the translocation stocks due to crossing-over, contamination of the strain by wild-type individuals, and partial sterility of the strain due to the translocation.

If no use can be found for the reared females, then it would be worthwhile to eliminate them before they consume expensive diet. A number of schemes have been proposed to accomplish removal of females at the egg or early larval stage. In some species, alleles resistant to specific toxic chemicals, such as ethyl alcohol, endrin, purine, potassium sorbate, dieldrin, cyromazine, and propoxur have been selected. Then translocations between one of the sex-chromosomes (usually the Y-chromosome) and the chromosome bearing the resistance locus are induced. When the toxic substance is introduced into the colony, only the sex carrying the resistant allele survives. This method has been successful for three species of mosquitoes and is being actively investigated for the medfly.

MOLECULAR GENETIC TECHNIQUES

The demonstration of genetic transformation of *D. melanogaster* with transposable vectors (P-elements) raised the hope that similar techniques could be employed in other insect species. At the least, the demonstration that exogenous DNA sequences can be introduced into *Drosophila* germ lines has suggested new approaches that can be taken with other insect species.

Unfortunately, attempts at using *Drosophila* P-elements as vectors of exogenous DNA have not yet been successful in non-drosophilids. Alternative gene vectors will have to be discovered. Some possibilities for vectors are various insect viruses, such as baculoviruses, which might form stable transformations. Investigators are also searching in other insect species for transposable elements analogous to the *Drosophila* P-element.

Some of the techniques developed in molecular genetics that may be useful in autocidal control programs for insect pests include: nucleic acid hybridisation, restriction mapping of chromosomes, nucleotide sequencing, cloning DNA sequences, insertion of recombinant DNA (transformation), and development of monoclonal antibodies.

Obstacles to progress in the use of molecular genetics for insect control are: 1) lack of a general vector for gene transfer, 2) lack of suitable selective criteria for useful loci, 3) uncertainty about what genetic modifications will prove most detrimental to a given species, 4) lack of knowledge of the basic genetic biology of most insect pests. None of these obstacles is insurmountable.

INTEGRATION OF PEST MANAGEMENT PROGRAMMES

The sterile insect release method and other autocidal control techniques are completely compatible with other types of insect control that might be used in IPM programs. In fact, E. F. Knipling has always insisted that autocidal control must be integrated with other measures in order to be used in the most effective way. The autocidal control technologies are most efficient and economical when pest populations are already at low levels. One reason for this is that these techniques generally involve release of laboratory reared insects. Thus, the numbers of insects that can be reared will determine the size of the area of release. If less native insects are in the release zone, then a higher release ratio will be attained or a larger release area can be covered.

Any insect control measure that will reduce population densities either before or during the application of an autocidal control program will enhance the effectiveness of both procedures. For example, use of biocontrol organisms is generally most effective at high pest densities, but lose efficiency as pest densities decrease. Thus, a Sterile Insect Release Program would be a natural adjunct to any type of biological control program. This same reasoning applies to the use of insecticides to reduce high pest populations and the use of pheromones in confusion programs.

Cultural control methods, which are generally used to reduce pest populations during their most vulnerable stages during the year, are completely compatible with the use of autocidal control programs because the latter programs would generally be used during the time of high reproductive potential of the target species. It is hard to think of an Integrated Pest Management Program that could not be enhanced by the addition of an autocidal control technique.

REFERENCES

Anonymous, "Radiation Induced F1 Sterility in Lepidoptera for Area-wide Control", *Proceedings of a Research Coordination Meeting*, Phoenix, AZ, International Atomic Energy Agency, Vienna, 1993.

Bartlett, A. C., "Insect sterility, insect genetics, and insect control", *In* D. Pimentel [ed.], *Handbook of Pest Management in Agriculture, Vol. II*. CRC Press, Boca Raton, FL, 1990.

Davidson, G., *Genetic Control of Insect Pests*, Academic Press, New York, NY, 1974.

Pal, R. and Whitten, M. J., *The Use of Genetics in Insect Control*, American Elsevier, New York, NY, 1974.

14

Integrated Pest Management

Three decades back, intensive and extensive cultivation of high yielding varieties of crops were introduced in India to increase food grain production. The new crop varieties and cropping sequences, especially monocropping, along with injudicious and indiscriminate use of pesticides created many problems. Several hitherto unimportant pest species began to cause economic losses. Others, especially polyphagous insects, have precipitated national problems. One glaring example is the old world bollworm, *Helicoverpa armigera* (Noctuidae: Lepidoptera), which caused a 66 per cent reduction in the yield of seed cotton in Andhra Pradesh State during the 1987-88 crop season. Previously, this pest only seriously affected rabi crops such as chickpea and was commonly known as the gram pod borer. Similar outbreaks have been reported for other insects: Whitefly in 1984-85 and 1985-86 in South Indian states, and tobacco caterpillar in 1977-78 and 1979-80 in the states of Tamil Nadu and Gujarat (Arunakumara 1995) and more recently, coconut mite, *Aceria guererronis* (Eriophyidae: Acarinae).

The agrochemical industry in India now produces 47,020 metric tons of pesticide (Directorate of Plant Protection and Quarantine 2002). Although pesticide consumption in India is low (around 500 g per ha) compared to other countries like Japan (12 kg per ha) and Germany (3 kg per ha), problems in India resulting from unregulated use are quite alarming. The predominant classes of pesticides used in India (during 2000-01) were insecticides, accounting for 61 per cent of total consumption, followed by fungicides (19 %) and herbicides (17 %). In India, most pesticide use is on cotton (45%), followed by rice (22%).

In recent years, low external inputs and traditional techniques, including non-chemical alternatives, have been increasingly urged for India. These are viewed as technology options that could help create sustainable systems and decrease or avoid the needs for expensive and undesirable chemical inputs. Alternative agriculture has argued for an economically viable production to be viewed in the context of a healthier, environmentally friendly and sustainable chemical agriculture and the need for investment in low external input and non-chemical alternatives that include farmer empowerment.

Numerous models exist and have been advocated and based on their respective inputs / nutrient management principles. They may be broadly classified into: (a) Integrated Pest Management (IPM), (b) Low External Input Sustainable Agriculture (LEISA), and (c) Organic Agriculture. Many sustainable agriculture initiatives and approaches based on these principles have now proved successful, with use of pesticides varying from more to only a limited amount. Of these, IPM is the one commonly advocated and widely adopted.

IPM IN INDIA

India has an agrarian economy, where the 1012.4 million population is dependent on agricultural commodities from 124.07 million hectares cropped area cultivated by 110.7 million producers. For rapid dissemination of IPM information, IPM related activities are being implemented through 26 Central Integrated Pest Management Centers (CIPMCs) located in 23 States and Union Territories.

Major activities under the IPM approach include undertaking sample roving surveys for monitoring pest/ disease situations on major crops, production and release of bio-control agents and conducting Farmers' Field Schools (FFSs). Pest/disease situations are monitored regularly in the states covering 644,000 hectares (ha) of the targeted area of 469,000 ha. Pest situation reports received from field stations and states were compiled and comprehensive weekly and monthly reports circulated to the concerned officers and scientists of State Departments of Agriculture/State Agricultural Universities and ICAR Institutes to help them take appropriate remedial measures.

A total of 16,260 thousand bio-control agents have been mass produced in the laboratories and released against insect pests in rice, cotton, sugarcane, pulses, vegetables and oilseeds (against the targeted release of 11,570 thousands during the year 2002-03. (40.54 % increase than expected) An area of 523 thousands ha has been covered against the targeted area of 367 thousands ha. in different states against various insect pests through augmentation and conservation.

During the eight-year period from 1994-95 to 2001-02, the government of India spent nearly Rs. 14,926 million for biocontrol of pests on different crops, covering a land area of 4.3 million hectares. In addition to this, Rs. 59 million was spent for pest monitoring. The largest states of India, Madhya Pradesh, Maharashtra, Uttar Pradesh, Andhra Pradesh, and Karnataka, rank highest in accomplishing IPM. A smaller state, Punjab has also done well in IPM. The disparity between different components of IPM accomplishment should be noted.

It can be also seen that the states that are progressing with IPM, are also ahead in using synthetic agrochemicals. *Bacillus thuringiensis* (Bt) and neem-based insecticides art the major bio pesticides consumed in India. India needs an improved and newer group of bio pesticides. The exponential rate of increase in bio pesticide consumption (together with reduced synthetic pesticide consumption) is a positive sign that India's agricultural community is becoming more concerned about the negative consequences of agrochemical usage.

Economics of IPM

To date, successful IPM prorammes have produced many benefits. These include (i) lower production costs (at farm level), (ii) enormous savings for governments from reduced pesticide imports and subsidies for pesticide use, (iii) reduced environmental pollution, particularly improved soil and water quality, (iv) reduced farmer and consumer risks from pesticide poisoning and related hazards, and (v) ecological sustainability by conserving natural enemy species, biodiversity, and genetic diversity.

The acceptability of a farming practice is primarily determined by short-term profitability. IPM attempts to integrate available pest control methods to achieve a farmer's most effective, economical, and sustainable combination for a particular local situation. Studies have been carried out to examine the resource use pattern and profitability of IPM vis-à-vis non-IPM farming practices.

Although IPM has been accepted in principle as the most attractive option for protection of agricultural crops from the ravages of insect and non-insect pests, implementation at farm level in India had been rather limited. Production uncertainty is commonly believed to be an impediment to adoption of less pesticide-intensive methods in agriculture such as IPM.

Integrated Pest Management (IPM) is uniformly recognized as one of the most useful and productive concepts in modern agriculture. Regretfully, however, IPM "implementation" often is cited as a pressing need to abate the perceived rampant use of pesticides. IPM is not - and never has been - a tool to reduce the use of pesticides. Rather, IPM is a rational approach to the incorporation of available economic and biologic data in an objective manner to assess the relative costs and benefits of pest control.

Qualifying Elements and By-Products of IPM

A precise and uniformly accepted definition of IPM has been elusive. Two key criteria, however, seem to prevail in most discussions of IPM:

1. Use of Objective Value-based Decision Making.
   — Control decisions should be based on maintaining pest populations below calculated EILs (as described previously).
2. Interactive use of Multiple Management Tactics
   — Diverse tactics such as natural enemies, host plant resistance, and pesticides are considered for use within a total management strategy

With real-world implementation of IPM using the above criteria, several by-products may emerge. Perhaps the most commonly touted by-product is the reduction in pesticide (or other tactic) inputs — and indeed this can and has happened in many instances. However, just as likely may be the by-product of more pesticide (or other tactic) inputs as a result of better understanding and realization of the injury (I) or damage (D) potential for pests. A third by-product, understanding the income risk that pests pose, will establish the "needs assessment" to objectively evaluate all tactical management alternatives. Although the by-products of IPM implementation can be significant - it is important to emphasize that they are by-products - not qualifying attributes. A valid IPM proramme does not have to decrease pesticide inputs to be an IPM proramme - indeed, the inputs may increase and still be considered IPM.

As a result of the confusion between qualifying elements and by-products, IPM is experiencing an Identity Crisis. Within academia and many branches of government, IPM has become somewhat of a "buzzword" used to convince granting agencies that their research proposals have merit as a means to the misguided goals of reducing pesticide use. And, although, many of the proposals do in fact relate to the key EIL biologic variables of injury (I) and damage (D), many more do not and contribute to the confusion and diffusion of IPM research objectives. In addition to academia, IPM also experiences a socio-political identity crisis - there is an errant belief that IPM is an "acceptable" public policy to reduce pesticides without the outright prohibition. Obviously, this identity is clearly wrong and not consistent with the management objectives or qualifying elements of the concept and practice.

IPM IMPLEMENTATION

In order to be truly sustainable, IPM must be supported as a sound financial enterprise represented through value-added services and products. Indeed, this is the most significant hurdle and places product and service organizations in the driver's seat with regard to accountability for IPM implementation.

At least four principles must be considered to support IPM implementation:

1. Value is the key concept of IPM, but can only be defined by the user.
   - We must recognize that each farmer will have a different business goal and therefore perspective on value-added IPM.
   - We must overcome our third party indifference to "income risk" as a key value-added quality of

certain control tactics within IPM

— We must develop techniques that account for value-sensitive differences among farmers.

2. Value will be exploited as a business enterprise.

— We must recognize that value-based IPM will develop sponsors and financial stakeholders.

— We must overcome the belief that IPM can be sustained as a public non-profit endeavor.

— We must develop business support mechanisms along with continuing education processes.

3. IPM is input-independent, only value should drive decisions.

— We must recognize that input analysis should be objective and based on financial criteria, but mediated by subjective valuation from the user.

— We must overcome the prejudices against any single tactic or class of tactics (e.g., pesticides) and the use of broad recommendations that do not consider individual grower goals.

— We must develop objective criteria for assessing input value (e.g., expand the use of EILs).

— Regulatory emphasis should be on real risks with specific tactics.

— We must recognize that regulatory policy should be invisible to IPM, with costs intrinsic to the costs of the control tactics.

— We must overcome the tendency to associate IPM with the promotion of biological or reduction in chemical use tactics.

— We must develop solidarity regarding risk assessment versus perception.

These principles - supported by a financially viable pest management industry (products and services) — will ensure sustained implementation of IPM. Without them, IPM is another short-term public initiative borne out of sociopolitical circumstances and artificially motivated through public grants.

FUTURE OF INTEGRATED PEST MANAGEMENT

As a sustainable approach combining biological, cultural, physical, and chemical tools to regulate pest populations while minimizing economic, environmental, and human health risks, IPM combines essential aspects of efficacy and safety to meet expectations of those who produce and market commodities and of the general public. The traditional focus of those involved with production of commodities has resulted in investment of tremendous resources for IPM in agriculture for research, educational endeavors, and for applications on-farm. It is becoming increasingly apparent, however, that principles of IPM have utility in virtually any situation where some species have been ascribed pest status. For example, in nursery and greenhouse ornamental production reliance solely on chemical pesticides to limit insect and disease problems is giving way to IPM.

In storage, processing, and distribution systems for cereal grains, unilateral reliance on chemicals for management of a variety of pests (insects, rodents, and microorganisms) is evolving toward integrated approaches with greater emphasis on sanitation, aeration of stored grains, even stocking plans for reducing infestations once packaged products have reached grocery store shelves.

The Food Quality Protection Act will require extensive education of those who work in food production, processing, and distribution systems regarding use of pesticides. New provisions relating to tolerances for pesticide residues in food commodities will necessitate planning throughout these systems to evaluate the potential cumulative human exposure to residues of certain classes of pesticides used in a variety of food products. This is just the type of comprehensive approach that will allow IPM practitioners to relate key food safety components to plans for managing pest populations as crops and livestock are raised, food commodities are processed, and products are distributed to consumers.

The manner in which food commodities are marketed has typically not been integrated into decision support systems for IPM The fact that commodity values may fluctuate by 50-100% depending upon the produce quality, detection of pesticide residues, and timely availability to meet demand must be considered as decisions pertaining to pest management are made. For example, market demand and value of fresh fruits and vegetables have reflected consumer concerns as indicated by increasing sales of organically-grown produce and that labeled as grown under IPM systems (northeastern U.S.). It is important that practitioners understand market trends and design pest management prorammes to assure quality, minimize pesticide residues, and preserve the identity of commodities as to production history and location. The following are factors that influence the marketability of food products:

— consumer demand for cosmetically-perfect produce without pesticide residues.

— concern of food processors relating to pesticide residues in fresh produce. (Over 50% of processors have changed purchasing contracts to minimize residues in their products).

— surveys indicate that over 50% of grocers will test fresh produce for pesticide residues, yet 98% indicate that they have not heard of IPM.

Economic considerations often rank foremost among factors that influence decisions relating to pest management. These considerations have been basic to IPM since its inception in the earliest definitions for the economic threshold and economic injury level. These parameters have not only served as criteria for decisions regarding the use of chemical pesticides in many plant and animal production systems, but have also helped to define the goal of effective pest management. Rather than an insistence on complete eradication of pests, an understanding is developing among practitioners of IPM that maintaining pest populations below the economic injury level by use of all available control methods is indeed successful control.

There remains, however, a perception that using economic thresholds as criteria for management decisions increases the risk of economic losses. The perceived risk is that sampling plans and decision criteria are inadequate with the result that pesticide applications may not be made when they are needed. This perception is often cited as a major limiting factor preventing the adoption of IPM.

While progress been made in defining simple economic thresholds and economic injury levels, especially for key insect pests, much research is needed for development of comprehensive thresholds for insect pests, pathogens, and weeds. Also, there has been relatively little accomplished in terms of integrating economic considerations relating to pest management activities into comprehensive benefit/cost analyses for agricultural enterprises.

Decision-making criteria for pest control activities in settings other than agricultural production have scarcely been addressed in research or extension efforts. While it is clear that the criteria applicable to urban settings will not be the same as those used in agriculture, value judgements should still be made as to the necessity for measures such as pesticides. It is of vital importance that these considerations be introduced to people in urban settings so that they have the opportunity to make informed choices regarding chemicals introduced into their environment and to develop an appreciation for the decisions that must be made in agriculture.

The scope of IPM implementation will continue to change greatly in the future. What began as limited approaches to promote integration of control methods for key pests and provide guidelines for decisions regarding pesticide applications is evolving to an ecologically-based systems approach. Where there was once limited cooperation among disciplines, now there are teams comprised of scientists from several disciplines that work with consultants and producers to incorporate effective pest management into production systems. The complexity of these systems requires a broad base of expertise to address issues such as timing and efficacy of pest control recommendations; dealing with the ever-growing problem of pesticide resistance; addressing issues of worker safety; avoiding pollution of soils and water by agrichemicals; and assuring that foods contain no harmful residues of pesticides.

While these issues must be dealt with for the forseeable future, it is important to assure an effective transition from pest control prorammes with heavy reliance on chemical pesticides to ecologically-based biointensive IPM. A recent report by the Consumers Union states that the way we control pest problems is at a crossroads with the necessity of decreasing reliance on chemical-dependent approaches becoming steadily more evident. An organizational strategy must be adopted to increase acceptance of integrated approaches by; 1) providing intensive educational outreach involving coordinated efforts of extension specialists, industry representatives, and private consultants, and 2) emphasizing regional and national coordination for preparation and dissemination of educational materials.

Those who develop and implement IPM technology must be proactive in design of information delivery systems. This challenge has several aspects including effective communication of research information to extension specialists to promote the formulation and delivery of integrated prorammes; providing timely advisories of the status of pest populations so essential to adoption of ecologically-based prorammes by IPM practitioners; and feedback to scientists from practitioners regarding effectiveness of prorammes. Also, the traditional flow of information from researcher, to extension specialist, to practitioner is being changed to include crop consultants and representatives of agribusiness.

REFERENCES

Bennett, G. W. & J. M. Owens, *Advances in urban pest management*, Van Nostrand Reinhold Co. New York, NY, 1986.

Cate, J. R. & M. K. Hinkle, *Integrated pest management: the path of a paradigm*, National Audubon Society. Washington, D. C., 1994.

Higley, L. G. & W. Wintersteen, "A new approach to environmental risk assessment of pesticides as a basis for incorporating environmental costs into economic injury levels", Am. Entomol., 1992.

Bibliography

Addor, R. W., "Insecticides", In: Godfrey C.R.A. (Ed) *Agrochemicals from natural products*, Marcel Dekker, Inc., 1995.

Anonymous, "Radiation Induced F1 Sterility in Lepidoptera for Area-wide Control", *Proceedings of a Research Coordination Meeting*, Phoenix, AZ, International Atomic Energy Agency, Vienna, 1993.

Armbrust, E. J. and G. G. Gyrisco, "Forage crops insect pest management", In R. L. Metcalf and W. H. Luckmann, *Introduction to Insect Pest Management*, John Wiley and Sons, New York, 1975.

Arnason, J. T., B. J. Philogene and P. Morand (EDS), *Insecticides of plant origin*, American Chemical Society, Washington, DC. USA, 1989.

Bartlett, A. C., "Insect sterility, insect genetics, and insect control", *In* D. Pimentel [ed.], *Handbook of Pest Management in Agriculture, Vol. II.* CRC Press, Boca Raton, FL, 1990.

Belderok, Bob & Hans Mesdag & Dingena A. Donner, *Bread-Making Quality of Wheat*, Springer, 2000.

Bennett, G. W. & J. M. Owens, *Advances in urban pest management*, Van Nostrand Reinhold Co. New York, NY, 1986.

Bryan, M. D., R. J. Dysart and T. L. Burger, "Releases of Introduced Parasites of the Alfalfa weevil in the United States, 1957-88", *USDA APHIS Misc. Publ. No. 1504*, 1993.

Buntin, G. D. , P. L. Bruckner and J. W. Johnson, "Management of Hessian fly (Diptera: Cecidomyiidae) in Georgia by delayed planting of winter wheat", *Journal of Economic Entomology*, 1990.

Caltigirone, L. E., "Landmark examples in classical biological control", *Ann. Rev. Entomol*, 1981.

Carpenter M.J, Ware G.W., *Defending Pesticides in Litigation*, 14th Ed., West Group, St. Paul, MN, 2004.

Cate, J. R. & M. K. Hinkle, *Integrated pest management: the path of a paradigm*, National Audubon Society. Washington, D. C., 1994.

Chang, T. T., *The origin, evolution, cultivation dissemination and diversification of Asian and African rices*, Euphytica, 1976.

Coats, J. R., "Risks from natural versus synthetic insecticides", *Annu. Rev. Entomol*, 1994.

Davidson, G., *Genetic Control of Insect Pests*, Academic Press, New York, NY, 1974.

Day, W. H., "Biological control of the alfalfa weevil in the Northeastern United States", *Chap 25 In. Papavizas, G. C. ed. Biological Control in Crop Protection*, Allanheld, Osmun & Co., Montclair, NJ, 1981.

De Datta, S. K., *Principles and Practices of Rice Production*, John Wiley, New York, 1981.

Debach, P., and D. Rosen, *Biological Control by Natural Enemies*, Cambridge Univ. Press, Cambridge, UK, 1991.

Donaldson D, Kiely T, Grube A., *Pesticides Industry Sales and Usage*, 1998 and 1999 Market Estimates. US EPA, Washington, DC, 2002.

Doutt, R. J., and J. Nakata, "The Rubus leafhopper and its egg parasitoid: an endemic biotic system useful in grape pest management", Environmental Entomology, 1973.

Ehrenfeld, D., *The arrogance of humanism*, Oxford University Press, Oxford, 1981.

Ehrlich, P. R., Wilson, E. O., "Biodiversity studies: science and policy", *Science*, 1991.

Eigenbrode, S. D., J. T. Trumble & R. A. Jones, "Resistance to beet armyworm (Spodoptera exigua [Hubner]), hemipterans, and Liriomyza spp. in Lycopersicon", *Journal of the American Society of Horticultural Science*, 1993.

Engelken, L. K., W. B. Showers and S. E. Taylor, "Weed management to minimize black cutworm (Lepidoptera: Noctuidae) damage in no-till corn", *Journal of Economic Entomology*, 1990.

FAO Report, "Sorghum and millets in human nutrition", 1995.

Grist, D. H., *Rice*, Fifth ed., Longman, London, 1975.

Heinrichs, E. A. and T. A. Miller, *Rice Insects: Management Strategies*, Springer-Verlag, New York, 1991.

Heinrichs, E. A., F. G. Medrano and H. R. Rapusas, "Genetic Evaluation for Insect Resistance in Rice", International Rice Research Institute, Los Baños, Philippines, 1985.

Herbicide Classification of the Weed Science Society of America, *Weed Technology*, 1997.

Higley, L. G. & W. Wintersteen, "A new approach to environmental risk assessment of pesticides as a basis for incorporating environmental costs into economic injury levels", Am. Entomol., 1992.

Janatuinen, E et al., "A Comparison of Diets with and without Oats in Adults with Celiac Disease", *New England Journal of Medicine*, 1995.

Pal, R. and Whitten, M. J., *The Use of Genetics in Insect Control*, American Elsevier, New York, NY, 1974.

Palmer, WE, Bromley, PT, and Brandenburg, RL. *Wildlife & pesticides - Peanuts*, North Carolina Cooperative Extension Service. Retrieved on 2007.

Shepard, B. M., A. T. Barrion and J. A. Litsinger, "Friends of the Rice Farmer: Helpful Insects, Spiders, and Pathogens", International Rice Research Institute, Los Baños, Philippines, 1987.

Smith, Albert E., *Handbook of Weed Management Systems*, Marcel Dekker, 1995.

Smith, C. Wayne, *Crop Production*, John Wiley and Sons, 1995.

Smith, R. D., "Background, use, and beneficts of blast freezers in the prevention and extermination of insects", en *Biodeterioration 6: papers presented at 6th Interational Biodeterioration Symposium, Washington DC, August 1984.*

Stranger, Carolla; Brandis, Leanne, "Insect pests and their erradication (non-chemical methods)", en *Australian Library Journal*, 1992.

Swaminathan, M. S., *Rice*, Scientific American, 1984.

Tribe, D., "Feeding the Greening World", *The role of international agricultural research*, CAB International, Wallingford, UK, 1994.

United Nations Food and Agriculture Organization: Sorghum and millets in human nutrition.

Watson, Andrew M., *Agricultural Innovation in the Early Islamic World: The Diffusion of Crops and Farming Techniques*, Cambridge: Cambridge University Press, 1983.

Wells M., "Vanishing bees threaten US crops" (in English), *www.bbc.co.uk, BBC News*. Retrieved on 2007.

Zhou, X., Jellen, E.N., Murphy, J.P., "Progenitor germplasm of domesticated hexaploid oat", *Crop science*, 1999.

Index